PRINCIPLES OF ELECTRONIC PRODUCTION

Principles of Electronic Music Production offers a comprehensive overview of the fundamental concepts that shape contemporary electronic music production, revealing it as a complex tradition.

Over ten chapters richly illustrated with the practices and ideas of innovative musicians past and present, across a range of styles, this book looks at the producer as an omnimusical figure—a composer-performer-engineer who improvises, designs sounds, programs MIDI, sculpts audio, makes beats, shapes arrangements, and balances mixes. The chapters explore the power of serendipities and musical systems, the phenomenology of listening, using minimalism and constraints, the lessons of AI, playing software as a musical instrument, the alchemy of sound design, creating dynamic musical form, and iterating tracks towards quality to answer a perennial production question: how does one make music that sounds alive at many levels at once?

Establishing foundations that music production students, professionals, and hobbyists can build on when thinking creatively in their own work, *Principles of Electronic Music Production* is a fascinating insight into the art of electronic musicianship.

Thomas Brett is a musician who has played percussion on Broadway since 1997 and holds a Ph.D. in Ethnomusicology from New York University. He is the author of *The Creative Electronic Music Producer* (Routledge, 2021) and writes about music at brett-works.com.

PERSPECTIVES ON MUSIC PRODUCTION

Series Editors: **Russ Hepworth-Sawyer**, *York St John University, UK*, **Jay Hodgson**, *Western University, Ontario, Canada*, and **Mark Marrington**, *York St John University, UK*

This series collects detailed and experientially informed considerations of record production from a multitude of perspectives, by authors working in a wide array of academic, creative and professional contexts. We solicit the perspectives of scholars of every disciplinary stripe, alongside recordists and recording musicians themselves, to provide a fully comprehensive analytic point-of-view on each component stage of music production. Each volume in the series thus focuses directly on a distinct stage of music production, from pre-production through recording (audio engineering), mixing, mastering, to marketing and promotions.

Remastering Music and Cultural Heritage
Case Studies from Iconic Original Recordings to Modern Remasters
Stephen Bruel

Innovation in Music
Adjusting Perspectives
Edited by Jan-Olof Gullö, Russ Hepworth-Sawyer, Dave Hook, Mark Marrington, Justin Paterson, and Rob Toulson

Innovation in Music
Innovation Pathways
Edited by Jan-Olof Gullö, Russ Hepworth-Sawyer, Dave Hook, Mark Marrington, Justin Paterson, and Rob Toulson

Music Production in the Music City
The Undersong of Place
Yanto Browning

Principles of Electronic Music Production
Thomas Brett

For more information about this series, please visit: www.routledge.com/Perspectives-on-Music-Production/book-series/POMP

PRINCIPLES OF ELECTRONIC MUSIC PRODUCTION

Thomas Brett

Routledge
Taylor & Francis Group
LONDON AND NEW YORK

Designed cover image: brizmaker via Getty Images

First published 2026
by Routledge
4 Park Square, Milton Park, Abingdon, Oxon OX14 4RN

and by Routledge
605 Third Avenue, New York, NY 10158

Routledge is an imprint of the Taylor & Francis Group, an informa business

British Library Cataloguing in Publication Data
A catalogue record for this book is available from the British Library

ISBN: 978-1-032-95491-2 (hbk)
ISBN: 978-1-032-95486-8 (pbk)
ISBN: 978-1-003-58514-5 (ebk)

DOI: 10.4324/9781003585145

Typeset in Sabon
by Taylor & Francis Books

CONTENTS

FIGURES

INTRODUCTION

Electronic music producers are seekers—ever in pursuit of new sounds, new sensations, and the techniques that make them possible. *Principles Of Electronic Music Production* summarizes the (current) state of this knowledge. It's intended as a compendium of ideas, strategies, techniques, and perceptions crowdsourced from talented musicians whose music and thoughts deserve amplification. Luckily for us, knowledge about electronic music production is more articulated than ever—what used to be tacit know-how has become widely shared discourse about best practices. This book is about the fundamental principles of electronic music production. It explains these principles by synthesizing perspectives and insights from the practices of some of the world's most innovative producers working in a range of musical styles. *Principles* is a companion to my previous work, *The Creative Electronic Music Producer*, which explored the steps of production workflows. This second volume continues the conversation, going deeper to understand production's skill sets and guiding concepts along a variety of thematic paths.

But why write about music production in an era of endless YouTube tutorials that promise hacks to master its every facet? One reason is that production is a complex, living musical tradition that deserves patient study. Producing requires numerous musical competencies and fluencies within a musical system. In a quest to create a dynamic music that feels alive at many levels at once, a producer engages with music from all sides—as a composer-performer-engineer, who writes parts, designs sounds, programs MIDI, sculpts audio, makes beats, automates effects, shapes arrangements, and finesses mixes.[1] In other words, there's *a lot* going on in production! So while one can't learn to produce solely by reading about it (or watching tutorials), this book takes a holistic view of the craft.

Writing is also a way to observe the themes and relationships that inform the production workflows in seemingly disparate musical idioms. Describing music production's landscape of possibilities organizes the most widespread, salient, and practically useful ideas into a map of a creative territory. *Principles* is not an argument, but rather testimony as to why these ideas about production are powerful.[2] Related to this, production's evolving practices are continually generating new ideas to write about. Each topic discussed in this book reflects practice and connects to a production principle. Practices

DOI: 10.4324/9781003585145-1

leave traces for us to explore, and by writing about these traces we reflect on how production skills illuminate production thinking. Lastly, *Principles* addresses recurring themes I've noticed over the years in online discussions about music production. On Reddit, for example, musicians ask questions:

What are sources of "creative" producing?
Is there a best workflow?
How do I work more effectively?
What is originality and how do I become original?
How do I develop musical taste and an instinct for quality?
What do great tracks have that mine don't?
Do production tools matter?
Can AI help me make music?
What makes a sound interesting and alive?

Principles tries to answer these questions and offers itself as a style-agnostic guide to the possibilities of production and strategies for implementing them. My hope is that musicians will find herein insights they can mull over and use in their own work. Each of the book's ten chapters explores different facets of electronic music production.

Chapter 1, "Serendipities", is about how producers set up musical systems with which to interact, curate and discover sounds, and configure their workflows as happy accident generators. Topics include organizing sounds, serendipities, adopting a musical *Zettelkasten* method, templates, working in abundance, exploration and exploitation, seeking beautiful outputs, and preparedness/randomness.

Chapter 2, "A Living Musical System (Alive at Many Levels at Once)", explores ideas and practices for interacting with one's musical system in a way that maximizes its creative possibilities. It examines the alive-at-many-levels-at-once concept, techniques for setting up a musical system, incorporating touch, populating sounds, playing an acoustic instrument versus playing a DAW (digital audio workstation), and deriving many musical lines from a single gesture.

Chapter 3, "Attention and Attunement", explores music production as phenomenology by considering how producers focus on perceptual minutiae in sound. It examines production as an attentional tool, hidden worlds in samples, the analytical-intuitive fader, play, musical process, getting each musical layer right, the untried, patience, and production-perceptual macros.

Chapter 4, "Minimalisms and Constraints", explores the power of minimalism, limitations, subtraction, and using the smallest viable number of parts to make a track. It examines the minimalist mindset and minimalist approaches to production, in-the-moment production constraints, process not outcome, the elegance of economy, and demo thinking.

The Interlude, "AI in Electronic Music Production", traces the arrival of AI in music production, from proto-AI composing techniques in the 20th century to today's text-to-audio practices and vocal cloning. The interlude considers the implications of prompt-based production practice, three case studies of producers collaborating with AI models, the state of AI voice modeling, and concludes with four AI-inspired lessons: the importance of taste, artistic influence as dataset, implementing musical prompts, and understanding creativity as error artifact.

Chapter 5, "Sonics and Instruments", explores ideas about sounds and instruments in electronic music production. It examines sound feels and vibe, instrument agnosticism, virtual versus acoustic instruments, stacking simple tools, ear candy/incidentals/subliminal little sounds, sonic artifacts, the appeal of analog sound, and the alchemy of imperfection.

Chapter 6, "Learning Producing and Building Tacit Knowledge", explores strategies for learning electronic music production as one practices it. It examines the concept of omnimusicality, finding one's sound, forward engineering, judgment and tacit knowledge, practice as repetition without repetition, repeated listenings, elements of production artistry, and error's opportunities.

Chapter 7, "Ambiences and Atmospheres", explores ambience and dimensionality in electronic music production. It examines ambient sound's history, producer's use of field recordings to escape flat virtual space, drone-making techniques, and the affect of effect chains.

Chapter 8, "Towards Dynamic Structures", explores how electronic music producers build cohesive tracks. It examines ways of using loops, grid musical thinking, tonal evolutions, musical form as story, and ways of developing ideas including recursive production, scaling, prolation canons, and transitions.

Chapter 9, "Iterating Towards a Quality Without a Name", explores how producers repeat the production process over and over to evolve and refine their music in pursuit of an aesthetic of energy. It examines iteration, tinkering, and the value of making a version 1.0.

Notes

1 As producer Brian Eno notes, "The idea of the producer as someone who mediates and converses between the completely non-technical musician and non-artistic engineer, which was the old picture, is now dead because most musicians now occupy all three of those roles" (Cunningham, 1994).

2 Here I take a cue about writing as a form of testimony from Verlyn Klinkenborg, who says that writing "shares your interest in what you've noticed. It reports on the nature of your attention. It suggests the possibilities of the world around you. The evidence of the world as it presents itself to you" (Klinkenborg, 2012: 117).

1

SERENDIPITIES

Introduction: organizing sounds to organize creativity

A few years ago on Twitter, the electronic music producer TJ Hertz (Objekt) asked the production community how they organize their audio samples:

> Producers: My sample libraries are a total rat's nest and finally need sorting this year. How do you organize yours? Do you separate samples from the pack they came from? Do you rename them for easier searching? Do you sort by category or by source? Help.
>
> *(Hertz, 2020)*

Hertz is a veteran musician with meticulous methods and years of experience, yet his sample library dilemma is every producer's dilemma in our time of sonic abundance. The fact is that inside the producer's primary tool, the DAW (digital audio workstation software), sounds are everywhere and easy to come by. It wasn't always like this. From the 1970s to the 1990s, a producer had fewer options, perhaps using a couple of hardware synthesizers, a sampler, a drum machine, and sequencing those sound sources on a processor-challenged computer. Back then, one's musical system and sounds were more limited than they are today. The upside to the state of the art at the time was that limitations of gear pushed producers to find innovative musical things to do with what they had.

Producers still seek innovative musical things to do, but the state of the art is different now. Music production involves repeated encounters with sonic abundance, and a vexing challenge is figuring out repeatable yet surprising ways to make sense of it. We need to understand sonic abundance's potential usefulness to our projects so we can get on with making music, and we need to find ways of interacting with our musical materials so that we're positioned to encounter interestingness. Hertz's question about how to sort his sample library is interesting because it's really about the organization of creativity in music production. This chapter explores how producers set up musical systems with which to interact and to curate and discover sounds, and configure their workflows as happy accident generators. Topics include organizing sounds, serendipities, adopting a

DOI: 10.4324/9781003585145-2

musical *Zettelkasten* method, templates, working in abundance, exploration and exploitation, seeking beautiful outputs, and preparedness/randomness. In general, the most useful sound organization system is one that allows for unexpected discoveries.

While a novice producer may feel lost amongst an abundance of sounds, experienced producers have ways of negotiating it. One way they do this is by leveraging their experience with sound to quickly get to what they need. Producers are metadata wizards, assessing and remembering sounds in terms of their style, mood, associations, and sonic features. Searching for just the right aesthetic feel for a track, producers sift through, experiment with, and finesse timbre variations of sounds. When they think, *I need a pad that's more liquid*, or, *this kick needs to be rounder*, they forge sound-to-metaphorical-descriptions-of-sound connections—as if searching through the metadata tags of sounds in all the musics they've ever listened to and made (cf. Walther-Hansen, 2020). Such thinking is quick and intuitive, a feeling-a-way-forward that guides them as they build and assess their track in progress. As we'll learn over the course of this book, experience is the most powerful production tool of all: its web of sound-to-feeling associations guides how we search for, organize, categorize, and work with the building blocks of music.

Back to Hertz's question about sorting samples. On its surface it captures a perennial production dilemma: *How should I sort my sounds to benefit from their abundance?*

Sorting usually involves placing the sounds in particular places (e.g. in a specific folder), or categorizing and tagging sounds by instrument, timbre quality, origin, mood, or some other criterion "to focus on certain properties that fit our purposes" (Lakoff and Johnson, 2008: 163). Consider two examples from the music software companies, Ableton and U-he. In the browser of Live 12, Ableton uses a descriptive sound-tagging organization system with two main parts, Sound (e.g. "Bass", "Lead", etc.) and Character (e.g. "Acoustic", "Synthetic", etc.).[1] The browser also features a neural network-powered "Sound Similarity Search" function which allows one to select a sample or preset and then "tell Live's browser to find comparable sounds and instrument presets" (Ableton, 2024a). In a similar way, U-he organizes the instrument and sounds in its synthesizers into Categories, Features, and Character.[2] Ableton's and U-he's systems of organizing and tagging are useful models for how to think about sound types, sound categories, and sound character (Figure 1.1). Whatever sorting system a producer relies on, the goal is to have one's sounds *findable when they're needed.*

Fellow producers answered Hertz's question with helpful tips from their own sound-organizing practice. The producer Evan Majumdar-Swift (aka 96 Back) suggested grouping miscellaneous drum machine samples (i.e. individual or "one-shot" drum hits) "into arbitrary folders like 'glass' or more ambiguously 'texture' then sub folders of a rough frequency range they inhabit." Another veteran producer, Joshua Eustis (aka Telefon Tel Aviv), suggested a single letter labeling scheme so that drums (D), keyboards (K), bass (B), and other sample types auto-sort themselves alphabetically. "Name like this: D Snare X, K Vangelis Lead C3, B Hoover C1…They will auto sort by name, folders not needed" (Hertz, 2020).

Such ways of organizing and naming help producers identify specific sounds so that they may be reliably retrieved at a later time. But sorting one's sounds so they're easy to find is secondary to a perhaps more interesting production goal: generating surprising juxtapositions with which to make music. We can call these surprising juxtapositions *sonic serendipities.*

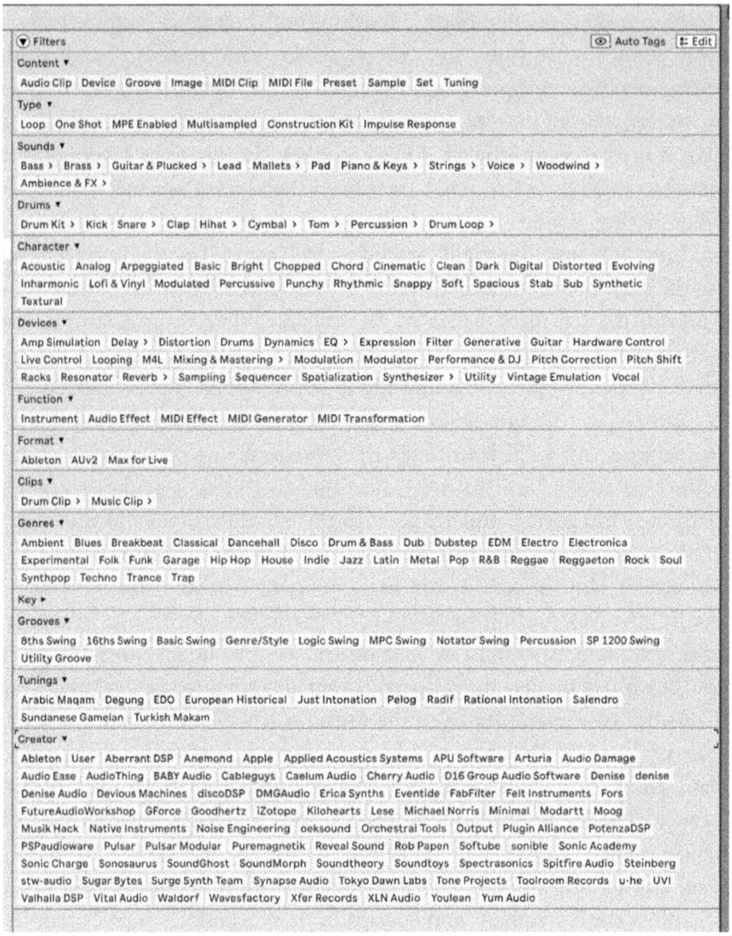

FIGURE 1.1 Organizing sonic abundance: Ableton Live's browser

Sonic serendipities

Serendipity is fortune's role in artistic and scientific practice via accidental discoveries of value. The term was coined by the writer Horace Walpole in 1754, with reference to an old Persian fairy tale, *The Three Princes Of Serendip*. In this tale, three perceptive Princes notice clues to insightfully describe a lost camel they've never seen. For Walpole, *serendipity* describes a way of being in the world, a pursuit of "always making discoveries, by accidents and sagacity, of things which [we] were not in quest of" (Remer, 1965: 6).

In electronic music production, making music feels most alive when it creates conditions for serendipities to emerge and for us to notice. We try to be perceptive, pursuing what we didn't know we were in quest of by listening for serendipities—in improvising, finding unexpected sounds and new uses for musical tools, answering the question *What to do next?* by adjusting timbres, juxtaposing loops, and recycling old work into new findings.

Sonic serendipities are cool sounds the producer stumbles upon accidentally. Producers often refer to these serendipities as *happy accidents* which were not foreseen. "It's the old hunt for surprises" says producer Thomas Fehlmann (Fischer, 2023). Examples abound. For producer Cristobal Tapia de Veer, improvisation is the main source of his accidental creative discoveries that come "from doing stuff, making errors and then realizing that something is amazing and following that path" (D'Silva, 2023). Or consider producer Alan Myson (aka Ital Tel), who describes an inspiring workflow occurrence: "I'll get inspired by some happy accident when I unplugged a guitar and it made a weird thud through the fx chain, then turn that into a percussive patch in Ableton" (Headphone Commute, 2018).

But a producer's stumbling upon is not entirely accidental. In the case of Myson's discovering a weird thud sound when he unplugged his guitar, we might say that the sound was co-created by an effects chain he had *already created* in his DAW. Or think of it this way: Myson's effects chain was waiting for its chance to trigger a sonic serendipity. So, the conditions for our stumbling upon are curated in advance—by us. "The challenge of any musical creation must be to invent the conditions of possibility for an appearing of the musical" notes composer and writer François J. Bonnet. "You don't create music; you create environments conducive to the advent of music" (Bonnet, 2020: 30). Sonic serendipities are a crucial part of music production's process because they tend to lead to cool-sounding results which, having caught our attention, may become interesting new ideas to work with—like Myson's percussive patch. For a producer, new ideas arise at the intersection of the organization of one's sounds and one's working within this environment.

Serendipity often appears when we're searching for one sound, then encounter another, more interesting one. Consider the seemingly simple task of locating a kick drum. If I know I want a Roland TR-808-style kick, I can find one, create one, or modify one. But it's often the case that I don't know exactly what kick I want until I try out a few and notice one in particular. Searching through samples or playing with parameters to alter an 808 sound I already have leads me into serendipity's realm. Now that I've happened upon an unanticipated kick drum that works, I wonder: *How many other interesting sounds await their discovery through a quick search or timbre tweak?*

We might notice that our sound searches are ad hoc and our default is to return to sounds—like an 808 kick—we already know where to find. Yet the convenience of this strategy forecloses new options. At every step of building a track through discovering and manipulating sounds, the producer is reminded of the complexities—the sheer combinatorial possibilities—of managing an abundance of musical information. Throughout a production's workflow, we're always recognizing that *this music could have been otherwise.*

And so, on a deeper level, Hertz' question about sorting one's sample libraries is about more than simply making sounds easy to access. Organizing sounds—knowing where they're located and how they may complement one another—is a fundamental skill that shapes subsequent production decisions and positions the producer to work creatively within the DAW's ecosystem of ever-changing sonic information. As Anders Reuter observes, electronic music production is "a new type of compositional practice based on controlling processual macro-synthesis of metadata rather than working with individual recorded sound sources" (Reuter, 2022: 2). In simpler terms, *the producer's practice is knowing how sounds can get along*. This is the knowledge of experienced producers who are metadata wizards because they can leverage their experience to navigate their soundsets. The artistry of music production is, at least in part, knowing how to deftly play with "the combination of digital information" (Reuter, 2022: 11).

In sum, it's worth a producer's time to continuously curate sounds by organizing and labelling libraries of samples, presets, instruments, effects chains, and so on. The goal is to get to know what you have and prioritize sounds and tools that speak to you. Even legendary producer Richard D. James (aka Aphex Twin), who once claimed to never save any sound he's already used, admitted that "I have a library of stuff, but it's stuff I haven't used yet" (Ward, 1993). What matters for any producer is getting to know the inspiring potentialities of one's musical system. By curating a sound world, a library of musical stuff, we set up conditions for discovery by making the most compelling bits available for our future selves to serendipitously encounter.

A serendipity machine: adapting the *Zettelkasten* method

> I don't know many people who have an endless stream of song ideas in their head. I think to be truly reliably creative you need to basically be able to generate some sort of stimulus for yourself to react to that is unexpected.
>
> *Mr. Bill*[3]

I want to suggest that when a producer organizes a library of samples, presets, instruments, and effects, the goal is not organization for its own sake, but rather a broader creative aim: to set up conditions for making beautiful discoveries. With this in mind, let's step out of the musical frame for a moment to consider electronic music production workflows from the perspective of a note-taking and knowledge management system known as the *Zettelkasten* or "slip box" method.

A slip box is a collection of notes written on index cards, each of which is tagged with numbers to connect it to other notes on other cards. The most famous advocate and practitioner was the German sociologist Niklas Luhmann (1927–1998). Luhmann built a slip box collection of some 90,000 cross-linked index cards and credited his practice with helping him generate ideas and write prolifically. In his visionary 1981 essay, *Communicating with Slip Boxes* Luhmann explains that the method takes a systems-based view of information and communication in which *meaning is generated within a system of connected ideas*. "Information is an intra-systematic event" he says at the outset. "It results when one compares one message or entry with regard to other possibilities" (Luhmann, 1981). In other words, information is a phenomenon of insight produced within a system.

Writers have long used index card files—paper-, and more recently, digitally-based—to keep track of their ideas and cross-reference them to produce new concepts. Vladimir Nabokov, for example, wrote character and story ideas for his novels on index cards, whose order he would shuffle to experiment with different plot structures.[4] But Luhmann's use of the slip box method was unique in that he added a numbering system for categorizing and, most importantly, *cross-referencing the notes in unusual ways*. The lesson for us is to think about the slip box as a serendipity-generating machine: a curated communication between the researcher/writer/artist and a system of idea storage, cross-connection, and retrieval intended to reveal surprising affinities and spark new ideas. The slip box method organizes creative work as a process of cultivating disorders, producing unanticipated juxtapositions, noticing cross-references and linkings, creating relations, and from this, producing insights and surprises.

In his essay, Luhmann identified slip box concepts that can help us understand where and how new musical ideas emerge. As we'll see, producers who develop and curate a generatively robust ecosystem of musical information with which to interact and respond are more likely to happen upon interesting ideas. Let's explore seven slip box concepts in the context of music production.

The musical index card

Just as the writer captures ideas on index cards, the producer captures sonic ideas in various forms and at various scales. The smallest unit of information is usually a sample, a loop, a chord progression, a melody, a synthesizer preset, or an effect. For most producers, a creatively rewarding practice over the long term is collecting such small-yet-interesting sounds, ideas, and presets to build a bespoke palette for future tracks. There are many ways to do this, but the essence of the practice is accumulating a library of one's sonic explorations, experiments, and findings. If something's interesting, save it! Consider six examples. Producer Robot Koch maintains a

> sample library which has a lot of my sound design experiments in it. Sometimes, I just create an interesting sounding thing and don't know what to do with it yet. It goes into my library and when I'm working on a piece of music I can just go through my library and see what fits in.
>
> *(Headphone Commute, 2022)*

Producers Marcus Eoin and Mike Sandison, of the duo Boards Of Canada, spend many years on each of their releases. They prepare for projects by recording all kinds of instruments into their samplers, processing the sound, and then resampling the results to make soundsets for their upcoming work. Eoin explains:

> So we'll maybe spend days just playing various things, wind instruments, strings, guitars, bass, synths, for hours into the samplers and then feeding those sounds through stacks of destructive hardware and resampling them to make unrecognizable new sounds. This is all before we even begin writing any tunes.
>
> *(Pareles, 2013)*

Producer and AV artist Halina Rice similarly curates what she calls a "sandbox" of sound design files to develop in future projects:

> Whenever I have downtime I create 'sandbox' files which are just for ideation. That might mean lots of time spent recording audio out of noise generators until there's a piece of sound that catches my imagination and I build on and around that. I often

> leave these files and come back later down the line when there's a project I'm working on—hearing with a fresh perspective—and working them up into finished projects.
>
> *(Headphone Commute, 2022a).*

Sean Booth and Rob Brown of Autechre are constantly developing their sound-generating Max patches and over decades these patches have evolved into what Booth calls "a pretty extensive deep system" (Muggs, 2016). He explains how returning to old patches allows him get reacquainted both with how they work and his own thinking at the time he built them:

> I quite like to build things and then forget how it works and then use it later on, and not really be able to remember what I was thinking about when I built it. It's a little bit like working with yourself in a way, but from a time when you're not aware of what you were thinking. You can get reacquainted with it in a sense, like you would with a person.
>
> *(Pareles, 2022)*

Producer Simon Green (aka Bonobo) has made recording-archiving percussion and Foley soundscapes a consistent part of his creative process since his first release, *Animal Magic* (2000). Green's bespoke sounds are either starting points for tracks, or a source of atmospheric sounds he can weave into his music as needed:

> I have an archive of drums that I've just collected—favorite kicks and snares and stuff over the years—but then I also layer stuff with a lot of Foley sounds, which I also collect. I've archived a lot of samples as I record them. So I'll have, for example, the noise that the little central console in my car makes when you close it. I've been sat there for hours thinking *that's a really good snare.*
>
> I record a bit of that and that's in the folder, and then little coffee cups that sound particularly percussive—*I'm just going to save that and go home and sample it so I have all those sounds ready to go.* And then more regular drum samples and conventional kick drums and hi-hats and everything else.
>
> *(Tape Notes, 2024)*

Perhaps the best exemplar of a musical index card maker is producer Kieran Hebden (aka Four Tet), who explains that his most powerful musical methodology has been to steadily curate an archive of audio samples. Hebden began his archive in the late 1990s and had the foresight and discipline to *keep adding sounds to the archive each month* ever since:

> Without thinking about it, the methodology I've applied has become more and more powerful over time. The idea was—and I started this in 1998—*I want to work primarily with audio, with samples and things.* Every month I just set up a folder on the computer and if I found something—be it listening to a record and there was a drum beat, or even if I was watching a film and there something on DVD or whatever that I liked I might record it into the computer. And I've just kept this and every single

> month there's a folder; it goes back to '98 and still goes on t'ill this day. It's just this sort of archive—things I've recorded on my telephone and stuff like that.
>
> *(Hanging out with Audiophiles, 2021)*

As Hebden's archive of interesting sounds grew over time, so too did its power to generate serendipities. In demarcating sound acquisition from the process of composing a track, the archive created the conditions for discovering what Hebden calls "interesting new bits" and answering the question of which sounds to work with. This freed Hebden to focus on his favorite compositional technique, *manipulating audio*:

> It started to become powerful after I'd had it for a little while, because when I'd sit down to make music instead of having that process being, *alright I've got to find some sounds that fit together, let me look for a record that's got this drum break on it, now let me pull out this keyboard and work on parts*. The minute I worked on music it was just me and the computer and nothing else. Because acquiring the sound had already been done. And then I started to find that the most original sounds I was making and the ones that were giving me sort of a signature style was when I was manipulating these samples to fit with each other. The artifacts and weirdness that would come out of that was proving to me more interesting than me picking up an instrument to play.
>
> A classic example for me is I play guitar: all the interesting guitar parts I was coming up with was me *messing with a sample of myself playing guitar*! It made me realize that me playing guitar wasn't that interesting to me anymore. Whereas me recording guitar into the computer and then a year and half later finding that file and having to distort it and squash it and manipulate it to fit with some other sample I'd found made all sorts of interesting new bits and gave the whole thing a sort of atmosphere. So that's been the way I've been working for years and it's still my favorite thing. The thing I keep coming back to is manipulating audio in a computer. That's where I seem to be able to make more magic happen than with anything else I mess around with.
>
> *(Hanging out with Audiophiles, 2021)*

Produce accidents and cultivate disorders

As much as we might want to be in control of the creative process, a producer's most powerful technique for making magic happen is finding ways to allow for aesthetically compelling accidents of timbre, rhythm, melodic/harmonic juxtaposition, and arrangement. As producer Jon Hopkins playfully approaches the matter, "Why not embrace and even cause accidents?" (Mackintosh, 2021). Or James Kirby (aka The Caretaker), who says "I want to create errors to see what can happen and to be surprised" (Bates, 2022). Or producer Fred Gibson (aka Fred Again), who describes "constantly trying to create as many vacancies as possible for accidents to happen" (Kamer, 2022). Accidents might transpire when we mess with a sample of ourselves playing an instrument, misalign two rhythms, invert, expand, or compress a melody into counterpoint, or loop three quarter's worth of a beat that makes an off-kilter groove. "The start of my creative process" says

TJ Hertz (aka Objekt), the producer whose question about how to organize one's samples opened this chapter, "is often about generating unpredictable results and finessing them from there" (Turner, 2022).

Producer Brian Eno was an early advocate of using the studio as a compositional environment in which to produce accidents and cultivate disorders (Cox and Warner, 2004: 127–130). Eno's approach was, and continues to be, primarily inspired by his interest in cybernetic systems. Thinking in terms of systems lead Eno to the idea that the producer's function is to create a musical machine that generates serendipities and then curate or—as Hertz says—finesse the results. In his book about Eno's landmark ambient album, *Music for Airports*, John T. Lysaker explains how the producer worked to initiate musical systems with which to interact through interventions:

> Taking his bearings from cybernetics, Eno works through interventions rather than utterly spontaneous creations. The variety he finds through systems music came from the system, after all. And therein lies the key—one takes steps to outwit oneself, whether with an oblique strategy or a pattern (or system) with which to generate sounds. That is, one initiates activities that run counter to one's habits, and one awaits the results, judgement at the ready.
>
> *(Lysaker, 2018: 101)*

Cross-reference ideas and concepts

Sometimes powerful composite sounds emerge when we juxtapose two or more contrasting elements. Consider, for example, the combination of bone-dry electronic percussion set against choral voices in a reverberant space.[5] The percussion feels futuristic, while the voices feel medieval. Through such sonic juxtaposition we sense ideas and associations from each sound world cross-referencing, meeting halfway to create a unique soundstage, to conjure a mood. Composer and organist Kali Malone speaks of the beauties of electronic-acoustic juxtapositions. "I love to combine synthesis and acoustic instruments in my work. There's an incalculable beauty to an acoustic timbre's organic quality and the human sensibility's delicate obscurity" (Bonnet, 2020a). Another example comes from producer Luke Pretty (aka Tennyson), who describes randomly choosing an initial soundset from archives of recordings, then mixing and matching its sounds to create surprising cross-references:

> I think the initial set of sounds is sort of just luck. With sound selection, I'm clicking in subfolders and dragging something random over. Having a lot of stuff to select from and being constantly surprised with what you're dragging over is helpful. I use [the peer-to-peer sharing network] Soulseek, where you can search generic classical music terms like 'largo quartet' or 'experimental 1980s' and just download 60 different recordings. So that's how I find stuff to use. But the way I'm mixing them is just completely random.
>
> *(Churchman, 2022)*

Discover preferred centers

As you work, it can be helpful to notice your preferred sounds and structures:

Are the sounds slow attack and soft-timbred, or hard-hitting and sharp-textured?
Are the sounds pristine or weathered?
Do they evoke the synthetic or the acoustic, the artificial or the natural?
Am I loop-oriented, or do I like longer, meandering phrases?

Notice the kinds of sounds you create and return to, or the presets you find compelling. Producers The Black Dog (Ken Downie, Martin and Richard Dust) assemble their preferred centers which they call *reference sources*. These sources are libraries of samples, loops, tones, and even video clips. They put these sounds into a folder to create a tone and style mood center for an upcoming project:

> When we start, we create simple banks of samples to try and get some very basic loops and tones going, just to try and set the tone and style that we want to start with. That usually gets completely thrown out the window by the end of the project, but because it's a start point, it helps us all to focus on a certain idea. We throw press cuttings into a folder, movies and YouTube links, so I guess we're creating an overall mood with everything that's available. It's not just Ableton project files we're throwing into the folders, but any kind of reference source we feel will be useful.
>
> *(Future Music, 2015)*

Discover linkings which result in the unexpected

As we layer and blend sounds in a track, we notice ways to relate them to one another, opportunities for connecting them. A chord part suggests a melody, two ambiances produce a composite atmosphere that blends elements of each, or two loops overlap into polyrhythm. Recognizing the linkings among our music's parts can lead us to develop these links and discover unexpected ways of developing the music, a topic of Chapter 2.

Discover ways of formulating problems so that heterogenous things relate to each other

Many of our cherished composite musical timbres relate well to one another because they've been used together for centuries. The orchestral string quartet, the brass section, the West African *jembe* and *jun-jun* drum ensemble, and jazz's piano-bass-drumset trio have histories of playing and repertoires that make their soundsets sound sensible. The composite sounds of these timbre orchestrations have existed over time as a unified force. But electronic music production, while a much younger tradition, has a significantly larger timbral palette with which to create novel sound worlds. Thus, the producer is always devising ways of relating unlike sounds—getting them to blend just so and sit well together in a mix. Recall Hebden working with samples of his guitar playing: his most original sounds came from *manipulating these samples to fit with each other*. The many other possible examples for relating unlike sounds—*Could this sample work with this saw wave? Could this beat accompany that amorphous resonance?*—get us thinking anew about how to build a track.

Cultivate combinatorial possibilities that were never anticipated

Cultivating combinatorial possibilities connects to producing accidents and cultivating disorders. By combinatorial we mean thinking about—and trying out!—combinations of a *limited* set of sounds, effects, and structures. For example, if our track has a pad, a bass, and a lead, in what ways can we swap these three sounds' roles, parts, and effects to create new combinations? Maybe derive a bass from the pad chords by muting all but their root notes and transposing these notes down three octaves. Or if the pad has a particular reverb on it, reuse a variation of that same reverb on the lead. Producers find ways to develop a track by reshuffling its existing elements rather than adding new ones. This creates a sonic world in which multiple sounds, as producer Austin Collins (aka Au5) notes, "come from the same family" (Brett, 2021: 98). As we'll explore in Chapter 8, there's an elegance about reshuffling and reusing work already in play. This is the essence of cultivating combinatorial possibilities: before introducing new parts or effects, recombine the sounds at hand.

In sum, these seven slip box-inspired concepts frame electronic music production in a way that shows the craft not unlike that of the writer who plays with ideas captured on note cards. When Niklas Luhmann noted that "information is an intra-systematic event" he meant that new ideas emerge within a system that has been deliberately set up to allow for cross-connecting smaller fragments into new patterns. So think about music production as a kaleidoscope for generating ideas. Try to apply slip-box concepts by organizing your practice in a way that leverages the musical equivalent of note-taking, accident-making, cross-referencing, discovering, relating, and curating the materials you already have to create conditions for serendipity to sing its song.

Templates, no template

A template is a starting point, a preset idea for possible musical directions. Since there are a vast number of ways we *could* go, using a template is a way to drastically reduce these ways to something manageable. A template is a collection of sounds, a *soundset*. A soundset is our bespoke palette, a toolkit of production possibilities, a collection of go-to enchanting instruments and timbres, a repertoire of effects and signal processing ways, articulated just so. A template's soundset helps solve the problem of how to begin today's work and redirects our attention from the abundance of *which sounds* to what shall I do with *these sounds*. For producer and educator Jono Buchanan, templates are idea machines, "an interesting way of being able to think about kickstarting ideas from a new starting point". Templates should be inspiring in a serendipitous kind of way. As Buchanan asks, "What if a template could produce some unpredictability?" (Buchanan, 2024).

But how do we decide what to put in our template? To answer this question in the context of making music in a DAW is to remember that a producer's template is an always-evolving arrangement of the most useful and inspiring sounds at the moment. A way to find a template's soundset is to try out various sound combinations until you develop a sense of what you like. While it takes time to get familiar with a soundset and there are always more soundsets to discover, it's helpful to commit to a few sounds that are exciting and powerful because they immediately conjure a world, a vibe. Soundsets are proxies for a mood we might not otherwise be able articulate, a voice that speaks on our behalf, ventriloquizing our feelings.

Of course, a template doesn't guarantee a musical direction. I've made templates with favorite instruments in them, but any plan I had for using these instruments dissolves once a project gets going. A common production experience is something unexpectedly stealing our attention. This reminds us how a single compelling sound—sometimes a sound *not* in our template—is enough to suggest a direction. So, knowing that my production plans often dissolve, I revise my template to suit each project. Sometimes I delete all but one or two instruments to keep my focus narrow. Other times the template includes a dozen synthesizers and samplers, ready to be serendipitously noticed. This is the approach of producer James Leyland Kirby (aka The Caretaker), who says, "I like having everything I can at hand and then knowing what will be best for a situation or work. I love the chaos of that, I love getting new sounds to use in wrong ways" (Bates, 2022). Inside each of my instruments are sounds that I've made, a topic we'll explore more in Chapter 2. But the instruments are also yet-to-be-explored universes of sound design—like templates within the template. The prospect of getting to know these possibilities is exciting, but also daunting because one only ever skims the surface of what an instrument can do. For this reason, perhaps the most useful template is one with just a few considered options.

If nothing else, templates help a producer sidestep the self-imposed pressure of beginning with the "right" sound. There is no right sound; moreover, there's no reason one's initial sounds need to be even close to right. For example, when building rhythm tracks, I sometimes default to TR-808 drum machine soundset to get a pattern going. Once I've sequenced a pattern, I explore ways of making it more unusual by processing its individual elements, a topic explained in Chapter 3. After a few rounds of processing the 808's sounds are unrecognizable. A template then, is just a self-imposed suggested starting point. There's no pressure to begin with the right sound when you understand that your process will, eventually, bring you to it.

Templates and optionality

In his book on risk, *On The Edge*, statistician Nate Silver shares lessons on how chance works from his experiences as a professional poker player. Silver understands our encounters with chance opportunities as "a reframing of optionality, which means making decisions that put you in a position where you'll have more choices to make later on" (Silver, 2024: 116). Optionality is apt for describing the function of a producer's template in that a template, by suggesting options, shapes how the producer will work.

To illustrate, over the years I've made many pad sounds in Xfer Records' Serum synthesizer, and these pads are what come to mind when I notice Serum in my template. I use pads as starting points for tracks, or sometimes as placeholder sounds to quickly record a part. But my options expand exponentially when my template combines different varieties of instruments. My pad sound is Serum, but bass tones may come from a Moog emulation, and marimba from a field recording. Suddenly I have a lot of options: encountering these contrasting sound sources inspires me to consider how timbres combine, how each one inhabits a different frequency and textural role. A Serum/Moog/marimba template suggests both options and directions. As I listen to these sounds as an ensemble, I sense potentials in their interactions that might steer me towards the next production move.

No template

Templates can point us in the direction of sounds, but they don't determine what we do—producers have no surefire templates for *how* to work. As producer Dylan Henner settles the matter, "I don't have a process" (Inner Islands, n.d.). Not having a process can be a good thing, for a reliable script for how to assemble a track can soon become a stale set of clichéd production moves repeated over and over. The danger, as producer and YouTuber Christian Brunn (aka Virtual Riot) observes, is that "if you want to make something *new* you might get stuck in your patterns of how to make sound" (Virtual Riot, 2023a). To avoid predictable results, some musicians consciously steer clear of pre-set ways of making music, aiming instead at ever-new workflows. For producer Barbara Braccini (aka Malibu), the aesthetic goal is admirably high. "Unless you invent a completely different way to make music, you're not making new music" she says (Bruce-Jones, 2023). Similarly, producer Nolan Petruska (aka Frequent) suggests that rather than use templates and preset sounds one should, ideally, *find a new way of making music on each project*. The downside of templates and presets, he says, is that they discourage the producer from thinking anew about each track's unique challenges:

> Templates and presets don't force you to think. And the less you think about what you're doing the more you end up in the same spot…For me it's *force yourself to find a new way, every time*. It's such a waste of time in a lot of ways, but I never feel like I'm making the same song twice, ever, because I have to do every bit of it manually.
> *(Live Producers Online, 2023).*

A purist who wishes to keep production's possibilities undefined might work without any template whatsoever. After all, an empty screen is pure possibility.

In sum, at the outset of a project producers follow their intuitions as they put together disparate sounds and textures for the first time. (Finer judgments and slow editing can come later.) Here a template helps steer our intuitions. It does this by setting parameters for our creative play within which we find a balance between doing compositionally sensible things and genuinely surprising ourselves by trying things that may seem strange, but could turn out to work. So even if, like Henner, you don't have a set way of producing, there may be techniques you rely on that consistently generate interesting results. Since techniques connect to templates, ask yourself, *Does my template support and enhance my workflow, or does it get in the way of trying new things?* Ideally there's a psychic-aesthetic fit between the configuration of your template and the idiosyncrasies of how your work. Stay nimble by adjusting and refining a template so that it feels inspiring, organized, and easy to use. Templates, no template: each time we produce music we recalibrate the equilibrium between working with sounds and instruments we've set up ahead of time and pursuing the unknown.

Working in abundance, exploration and exploitation

As we learned at the outset of this chapter, electronic music producers work with, and despite, sonic abundance. In a DAW-based software ecosystem, a producer encounters an abundance of sounds, known and unknown. Some of these sounds we haven't encountered yet, some of them we've encountered but forgotten where they reside, and some of them we've yet to create. Working with an abundance of sounds is both the pro and con of 21st-century music making using the computer as one's primary instrument: the options are vast, but we need to find ways of constraining this vastness to get something done. *Where to begin?* Do we go with something new, or revisit something already known?

The producer's abundance brings to mind the exploration/exploitation model of creativity, which has been used to explain "hot streaks" in artistic and scientific discovery (Liu et al, 2021). According to this model, exploration is an experimental tinkering around with new ideas to see what happens and what works, while exploitation is reaping the fruits of this exploration towards a specific goal by "focusing on a narrow area to develop deep expertise" (Liu et al, 2021). Exploration is speculative and a potentially risky use of time because it may never lead us anywhere significant, while exploitation is a more conservative practice because it tends to repeat what has already proved successful.

For the producer, exploration is getting to know the abundance of sounds one has access to. This process can feel laborious and slow. We can think about exploration as an ongoing, yet cumulative investigation of sounds we locate or create, making note of them (mentally and actually), and trying them out in various combinations. For producer Chris Clark (aka Clark), exploration happens through experimentation, out of which music eventually emerges:

> I find that abstract realm of experimentation—it's such a great place to start to just experience ultimate freedom. You're just in this zone of just dabbling...I just start there. And then the songs come out of that.
>
> *(RA Exchange, 2022)*

Producer Mark Fell considers exploration to be the essence of creative practice. Fell situates the composer's work in getting to know a set of sonic materials, rather than external inspiration:

> When creative practice is understood as an attempt to explore how materials and processes interact under certain conditions, like some scientific activities, it becomes situated in something—materials, processes, conditions—rather than somehow impossibly floating outside the material world.
>
> *(Fell, 2022: 21)*

Producer Ryan Lee West (aka Rival Consoles) spends most of his time in exploration mode, building musical things and trying them out in a kind of production-as-research process Fell would appreciate:

> For a long time I've got in the habit of just making—even when I'm not in the mood. And it can be anything. It can be fleshing out a composition in a half an hour, concentrating on an effect, concentrating on a melody—all the different aspects of making music. And then maybe that can be re-used a month later, or a year later, or never. It's almost like research for me. I'm constantly exploring stuff on different scales and then the stuff that interests me ends up on different compositions, potentially. But even if things don't get used, often you learn from them, or you learn what you like or don't like [of] the process.
>
> *(Why We Bleep, 2023)*

And producer Anna-Karin Berglund (aka AKB) sees exploration as introducing randomness, which keeps the writing process open-ended and helps her avoid getting stuck working on a track in a linear way: "To not work in a linear way when making a track. It's better to just start somewhere and explore from there, don't try and write a song from start to finish, make it random" (Moller, 2020).

Case study: "Slowdown"

A few years ago I revisited my track "Slowdown" (2022) to see how I might improve it before release.[6] Looking at the file I see that the track's primary sounds are pads I had created with Serum. During a few weeks of sound design, I had saved thirty variants of these pad sounds and made numerous tracks with them, as a daily practice. The structure of "Slowdown" is a 12-chord sequence played by one pad, overlaid with second part's responding chords on a different pad. Two accompanying percussion parts (made from samples of rolling balls) fill out the mix. The music unfolds through the dovetailing chords, slinking percussion, slow-building effects, and an atmospheric track derived from resampling the pad parts.

As I listened, I wondered how to improve the music by exploring and exploiting its components. First I noticed the effects: the pads have reverb on them, but what's its source? Once I located the reverb, I wanted to know what was making its sound wobble. I dove back into exploration mode: *Is it the Wow and Flutter controls? The Color control?* I twist knobs to reacquaint myself with the reverb, making small adjustments to make it more obvious. Now I can hear more details in the details. Exploitation in the context of "Slowdown" meant not messing with what I liked about the music in the first place, so I didn't change the chords or alter the sound design.

The second thing I noticed was that the piece clearly needed a bass, and the bass couldn't sound added after the fact—it had to fit. Back into exploration mode I went—making a bass from the chords, trying out bass sounds, and adjusting filter settings. Exploiting this bass part meant that its sound would have to be finessed further: its highest notes were jumping out, so I reduced their volume; the lowest notes were a tad lost in the texture, so I EQ-boosted them. I opened a filter wider, noticed how this made the notes-jumping-out problem worse, and closed it again, but not as much as before. Around and around I went, finessing minutiae of a few notes until the part blended in. I could keep adding to "Slowdown" but it seems to have enough. Producers tread carefully in sonic abundance as they usher their work along, using the minimum of what the music needs to thrive.

Seeking beautiful outputs

> I'm lazy; that's why I like machines. They do things I would not have thought of. I can put things into them, and then I can see something happen there beyond what I would have had the time, the taste, or the endurance to have produced myself. I usually don't want to slavishly make something in detail. I want to produce the conditions from which it and many its could come into existence. I think of myself as a machine builder in a way. Making a record for me is inventing a way of making music.
>
> *Brian Eno*[7]

Since making music in the DAW involves negotiating vast sound possibilities, the producer is always wondering, *What should I do now and how should I do it?* When we're unsure about how to develop a track, we would do well to consider a phrase used by essayist Naval Ravikant: *seek beautiful outputs*. Echoing Eno's thinking about generative musical systems, a topic we'll explore more in Chapter 2, we can understand production workflows as *devising conditions that might lead to beautiful outputs*.

Beauty is not the only musical goal—some producers prioritize complicated or intricate constructions, while others pursue simple, tuneful melodies, melancholy moods, shocking noise-scapes, *wub wub* bass, or propulsive, slamming beats. But beauty, however defined in whatever musical idiom you work in, is always part of the equation. And beauty slams by subtle means through all the small differences that combine to make a big difference. Beauty is that quality without a name we notice that elevates a track into something special, into something deserving our attention.

Compositionally, beauty in music emerges in many ways. It can be subtractive or additive, the by-product of what you take away from, or add to, your track. Or it can be cumulative, the result of multiple parts layered and interacting. It can also be accidental and beguiling, like when you serendipitously find or make a sound you didn't know would work until you tried it. Physicist Michio Kaku offers a compelling explanation of beauty that illustrates the kinds of outputs music producers seek. Kaku defines beauty in terms of *symmetry* and the re-arrangement or rotation of an object's or phenomenon's components to create novel, self-similar patterns:

> Beauty to a physicist is symmetry. There's a symmetry in music. For example, the simplest symmetry is a rubber ball. You rotate the ball and it remains the same. Why is a kaleidoscope beautiful? A kaleidoscope is beautiful because you rotate it and it turns into itself. Why is an ice crystal beautiful? Because you rotate an ice crystal by sixty degrees and it rotates into itself. So that's what beauty is: if I re-arrange the components of an object, it remains the same. Now you can apply that to music, you can apply that to physics. When you apply that to physics, it means that I have an equation and I rotate its components in a certain and precise way and it rotates into itself. That's called symmetry.
> *(Ferriss, 2022)*

Ravikant's idea of seeking beautiful outputs and Kaku's idea of beauty as symmetry are guides to music production. As you gain experience making tracks, you begin thinking about a music's early stages as *inputs for its latter stage outputs*. Improvising is a good example of an early-stage input. If you're a keyboard player, your improvisations are inputs for other parts not yet created. Scale back what you play: use fewer notes in your chords, or leave more space in your playing. As Nate Silver would say, your musical decisions *position you where you'll have more choices to make later on*. Later in the production process you can fill these chordal and temporal spaces with other sounds.

We can scale back not just our improvisations, but also our expectations. If we think of improvising as input rather than a finished piece, our playing doesn't need to be perfect. It just needs to be an interesting starting point. What makes an improvisation interesting? It should hold your attention for its duration. If you're not feeling it, start again. Repeat the process until something's happening, until the improvisation has life—as a solo, as a little journey free of supporting parts, effects, or production tricks. Would you listen to this as it is? If yes, move on.

With an improvisation recorded you can switch gears, reassured that you have inputs to play with, inputs whose elements you can rotate and re-arrange—like Kaku's kaleidoscope and ice crystals—as you work towards beauty. The producer Gábor Lázár describes his initial musical ideas as input for an environment in which he's working, while his output is the result of interacting with production tools he uses to alter the ideas:

> Of course I have some kind of idea, but many times I just look at my idea as a starting point and then I let the technology change it. So basically my idea is an input for the environment in which I'm working, and then there is an output which carries out the idea and the properties of the technology itself.
>
> *(Ravens, 2020)*

Trust that your improvisation can spawn beautiful outputs. There are reliable techniques that producers use to make this happen, some of which are explored elsewhere in this book and all of which are worth learning—

resample it,
cut it into pieces,
re-pitch it,
put it into a sequencer,
effect it into another sound,
convert the MIDI to audio and the audio to MIDI,
stretch and slow it down,
compress and speed it up

—but perhaps the most exciting way to discover beautiful outputs is by trying different techniques for doing so, each time you work. This is what producer Nolan Petruska (aka Frequent) meant by forcing oneself *to find a new way, every time.*

Just as your improvisation should hold your attention, so too should your workflow. So keep it interesting by doing something new *each time around a process to turn repetition into practice.* For example, chaining effects together, a topic discussed in Chapter 7, is a simple yet reliable way to achieve original sounds. Keep changing effects settings or swap in other effects until *something is happening.* Even the smallest changes to an effect's settings can produce massive output differences. Save your effect chains and their variations, only re-using them once you've altered them again. In production, nostalgia for what we've already done can block us from exploring the yet to be. Be bold by being resourceful and tinkering with what you already have. In sum, the techniques a producer uses are less important than whether or not they create outputs that are in some way beautiful. Computers, software, and sounds come and go. But a beautiful output is always a prospect on the horizon, beckoning and serendipitously within reach of our imperfect practice.

Preparedness and randomness

In addition to working in abundance and seeking beautiful outputs, producers curate serendipities. Building on our earlier discussion about organizing sounds and adopting *Zettelkasten*-inspired methods, we curate serendipities by positioning ourselves to encounter interestingness. Now let's dive a little deeper into the matter. The way we encounter interestingness is by spending time at the intersection of *preparedness* and *randomness*.

Preparedness is being intentional. It involves getting to know our tools for making music—creating, altering, and saving sounds, effects and chains of effects, sifting through and evaluating pieces in progress, and learning, day by day and bit by bit, the affordances of our musical system. Getting to know our tools gets us thinking, *What might I do with them?* The more we know what we *can* do with a tool, the more ideas we have about what we *might* do. Moreover, spending time intentionally learning our musical system accelerates our learning about production, a topic of Chapter 6. Learning, as computer scientist and essayist Paul Graham notes, is *superlinear* in that it grows exponentially. "The more you know, the easier it is to learn new things" (Graham, 2023b).

Reencountering previously made sounds or effects can be serendipitous, but this serendipity depends on what entomologist Fredrik Sjöberg calls *limiting our field of search* to what we've already made or altered for the purpose of being findable.[8] This idea of having one's previous work ready to go leads us to a strategy within the serendipity production principle: *position yourself to encounter interesting sounds.*

If preparedness is about deliberate positioning, chance is the random X factor that brings us beyond our intentions and shapes our encounters with the sounds and tools we've prepared. Designer and writer Leonard Koren speaks of the value of surrendering to chance and committing to its results in a disciplined way:

> Sometimes we don't want to make decisions. It's not that we can't or are unable to but rather desire that our next move come from a place beyond our usual controlling mind with its predictable prejudices and tastes. Perhaps arbitrariness, the wisdom of the universe as reflected in the random workings of chance, is the answer...Of course, we also have to commit to follow wherever chance leads us, even if that seems unappealing at the moment before we make our move. Otherwise we're back to same old same old.
>
> *(Koren, 2023: 57)*

Sound artist Ryoji Ikeda, who composes large-scale digital video and sound installations using a minimalist's soundset of sine tones in extreme registers, speaks of balancing randomness and control, finding ways for the two forces to exist in counterpoint:

> I try to control randomness. This is a big counterpoint, the encounter of randomness and control. The contrast is more interesting. If you really control a millisecond, there are other possibilities, even if they are microscopic. You can't perceive the change directly, but if you pay very close attention, the entire composition changes. So, I try to add randomness, and I like to see the counterpoint, the counterbalance.
>
> *(Weibel, 2014)*

Producer Joe Powers (aka Proc Fiskal) exploits randomness through a technique he devised for "playing" the timeline of YouTube videos whose audio he uses as source material for tracks. Powers plays YouTube as a sample instrument that randomly triggers the start-points of samples:

> I sort of play YouTube like an instrument and put things straight into Audacity [software]. If you play something on YouTube, you can hit the number key....4...7...2 or whatever and it skips the video so you get a sort of random sample allocation... Tracks came about just by doing that and mashing the random start-points in a video.
>
> *(Mackintosh, 2022)*

And producer Daniel Lanois has devised a system of sampling from a track's existing parts, processing them, then incorporating the "doctored" material back into the music as a texture. While Lanois deliberately prepares the system, he runs it randomly from time to time, just to hear what happens:

> I developed this system of taking samples of already existing components and extracting them from—putting them out of sync with the track and then doctoring them externally through other boxes, maybe changing them to slow them down, and if I hit on something special then I go back in and find a spot for that special sound back into the track. It won't work for most of the songs. I just run it randomly.
>
> *(Graham, 2016)*

Producers encounter randomness whenever they interact with the sounds and effects they've made, altered, and saved. The lesson here is to keep trying out sounds until something jumps out at you. When a sound does jump out it's an opportunity to quickly fine tune it, re-name it, save it, and most importantly, *use* it! Fine tuning-naming-saving-using allows the producer to step out of the compositional frame for a moment to add to their ever-growing repertoire of tools which may prove useful on another track. To echo what producer Ryan Lee West (aka Rival Consoles) noted earlier: even if we never use the sound again, we've learned something from our random encounter with it.

Tinkering with sounds that may have no immediate utility and saving them for later use is ultimately a way of cultivating a garden of what composer Harold Budd called "closest at hand" sounds that can be re-stumbled upon (Brett, 2021: 34). The power of closest at hand sounds is that they're pre-filtered by your former self. They give you license *not* to search endlessly through the DAW's abundance of sounds with little connection to your work. Closest at hand sounds limit our field of search which in turn keeps our workflows flowing. In this way, production is a species of gardening, where sounds are sown and reaped in an ongoing creative cycle of exploration and exploitation.

In sum, the intersection of preparedness and randomness is compelling because there are so many surprising outcomes that might result. We conclude with just a few examples:

- You try out an effect you made for one type of sound (say, percussion) on another (say, voices) and it's spectacular. You hadn't anticipated this sound, but now it's an option.
- Even though you've made many sounds, nothing fits the current project. This momentary dead end may indicate a mismatch between the sounds you have and the sound you need.
- You randomly open an instrument you haven't used in a while, discover a sound you'd made but forgotten, and it fits the track. What are the chances of *that* happening?

Notes

1 Ableton's Sound types are: *bass, brass, guitar & plucked, lead, mallets, pad, piano & keys, strings, voice*, and *woodwind*. Character types are: *acoustic, analog, arpeggiated, basic, bright, chord, cinematic, dark, digital, distorted, evolving, inharmonic, lofi & vinyl, modulated, percussive, punchy, rhythmic, snappy, soft, spacious, stab, sub, synthetic*, and *textural*.
2 U-he's Category types are: *bass, pads, leads, keys, fx, drums, seq + arp*, and *other*. Features types are: *mono, poly, duo, chord, BPM, modulated, dry, percussive, soft attack, slow release, glide*, and *wavetable*. Character types are: *bright, dark, consonant, moving, clean, dirty, soft, aggressive, phat, thin, natural*, and *synthetic*.
3 The Mr. Bill Podcast (2022).
4 Nabokov explains how he uses index cards to chart a story's structure: "The pattern of the thing precedes the thing. I fill in the gaps of the crossword at any spot I happen to choose. These bits I write on index cards until the novel is done" (Gold, 1967).
5 I used this percussion and voices juxtaposition on my 2022 recording, *Cadences II*.
6 The track is available on Spotify.
7 (Kelly, 1995).
8 Sjöberg writes: "Although there is much in this world that is incomprehensible, you can nevertheless discover a meaning as long as you have managed to limit your field of search" (Sjöberg, 2017: 167).

2

A LIVING MUSICAL SYSTEM (ALIVE AT MANY LEVELS AT ONCE)

Introduction

This chapter explores ideas and techniques for interacting with an electronic music production system in a way that maximizes its generative/creative potential. It examines configuring a musical system, hearing the human touch, populating sounds, playing a DAW as an instrument, and deriving many musical lives from a single musical gesture. I explain how producers work in a feedback loop with their bespoke musical systems, proceed intuitively, and use judgment while building tracks.

In his prescient 1968 article on electronic music, critic Carter Harman noticed and described its machine-assisted clichés—its "bleeps and chirrups, along with that robotic reverberation and that inhumanely precise reiteration" (Harman, 1968). But Harmon also heard beyond these clichés to recognize the electronic music studio as "the most important new instrument since the piano." The studio, he said, offers

> a beguiling array of possibilities for the exercise of fantasy in tone colors and textures, a microscopically subtle control of pitch and rhythm, and the opportunity for virtuosity and complexity in quantities limited only by the composer's imagination and patience.
>
> *(Harman, 1968)*

Today, electronic music producers' studios are their musical systems. No matter what kind of music you produce, you work with some type of system that helps you generate *a beguiling array of possibilities* for compelling and enchanting sounds. We can think of a musical system as a cybernetic circuit that connects the musician with a set of tools. These tools can be software, hardware, acoustic instruments, or a mix of all of these sound sources. It's common for producers' systems to revolve around the components of a computer-based software studio, commonly referred to as a DAW or digital audio workstation.[1] In this scenario, the system is the DAW + its sounds, plug-in instruments

DOI: 10.4324/9781003585145-3

and effects, and you, their collaborator. A musical system ideally feels full of potential insofar that it can generate serendipities, multiply possibilities, and co-produce cool ideas for us to develop. Like a game of Pong, the longer we play a musical system, the quicker it bounces output back at us. A well-configured musical system accelerates a producer's thinking about what's sonically possible.

The simplest way for a producer to build a musical system is to work with a single instrument, sound, or device. Play with it, explore its parameters, experiment to hear what it can do. As you discover sounds you like, build music with them. Then find ways of relating and connecting the sounds to those of other instruments and effects devices. Over time you build a world—a production ecosystem in which to work. Richard James (aka Aphex Twin) puts it thusly, "You need to think to set things up then you want to channel whatever it is from wherever it's coming from" (Nosnitsky, 2018). The goal is to get the components of the musical system *interacting* with one another—to use the system to build a circuit of connections through which possibilities flow. As producer Kirk Thompson (aka Krust) recalls about the pioneering drum and bass music he and his colleagues made in the early 1990s using the DAW Cubase, an Emu sampler, and a mixing board:

> What we created was a system of making jungle music. We made a *system*—in the arrangements, in the sound, in the way beats worked. We really dug deep and figured out how to make this sound work as a *sound*.
>
> *(SOS Electronic Music, 2024)*

Alive at many levels at once

The point of a musical system is to help the producer create exciting music—music that feels *alive at many levels at once*. This is an important idea because it characterizes what makes any music—from dronescapes to dance tracks—an experience that feels captivating, animate, and immersive. Music that's alive at many levels at once radiates presence, vitality, variation, interaction, responsiveness, suggestion, hidden depths, and multiple elements working together to create synergy.

The alive-at-many-levels-at-once concept was coined by architect and design theorist Christopher Alexander in his 1979 book, *The Timeless Way Of Building*. The phrase describes an ecosystem as a connected world constituted by interdependent patterns. The by-product of these patterns is an ineffable, yet perceivable sense of *a quality without a name*. Alexander vividly illustrates the concept with a beach scene (Figure 2.1):

> The quality without a name occurs, not when an isolated pattern occurs, but when an entire system of patterns, interdependent, at many levels, is all stable and alive.
>
> We may see the sand ripples anywhere where we choose to put loose sand under the wind. But when the wind blows across the sea, over the inland marshes, and the sand ripples support the dunes between the two, and the sandpiper walks out, the sand fleas hop, and shifting of the dunes is held in check by grasses which maintain themselves and the sandpiper—then we have a portion of the world, *alive at many levels at once*, beginning to have the quality without a name.
>
> *(Alexander, 1979:131, italics added)*

FIGURE 2.1 Alive at many levels at once: sea, sand, wind, and dunes

From the beach to the studio: Alexander's scene, in which the wind blows across the sea, marshes, sand dunes, and insects to form *an entire system of patterns*, inspires us to imagine and make music that is similarly alive at many levels at once, moving like nature does, interconnected and vital with interpenetrating energies.

Perhaps such talk sounds grandiose, yet producers often speak of techniques they use to make their tracks behave like living systems. Producer Brian Eno, who we discussed in Chapter 1, was one of the earliest champions of this approach to workflows. Eno was inspired by the cybernetic/systems theories of Stafford Beer whose thinking is summed up in a simple phrase he often used: *the purpose of a system is what it does*. Eno explains his interest in how a producer can "ride on" the dynamics of a system that is only loosely specified:

> Cybernetician Stafford Beer had a great phrase that I lived by for years: Instead of trying to specify the system in full detail, specify it only somewhat. You then ride on the dynamics of the system in the direction you want to go.
>
> *(Kelly, 1995)*

Eno's *Music for Airports* illustrates this *ride on the dynamics of a musical system* idea. The system Eno used to compose the piece comprised tape loops of varying lengths. Because their lengths were not multiples of one another, the loops never overlapped the same way twice. *Airport*'s musical system thus produced a continually evolving kaleidoscope of shifting tones and harmonies. In the decades since Eno's early ambient experiments, producers have continued to pursue ways of configuring their musical systems so that sounds behave as if alive at many levels at once. Let's explore some examples.

A common configuration is one whose elements are continually in motion, yet remain, like Eno's tape experiments, steerable. For example, producer James Holden explains his interest in finding ways of making sequences that don't sound machine-made. Instead he seeks "things with that level of movement and complexity and expression but not as predictable, not as linear" (Scarth, 2013). Holden's approach to production "is to try and set [a track] up as a living system where everything's moving by itself a little bit, and interacting with each other, but I can steer it" (Mullen, 2023). One way he achieves this is by configuring sequences whose rhythms and pitch sets are independent of one another:

> I guess most of the important stuff is my sequencers. I try to program sequencers that are separate from the scale...so the sequence has a shape and a rhythm in a scale, and then you manipulate those things separately. You can transpose it, but in the scale, or change the shape or add ornaments and stuff to it.
>
> *(Mullen, 2023)*

Similarly, producer Boris Potschubay (aka Jogging House) is interested in sounds that ebb, flow, and change like waves of water do:

> I do like sounds that are like waves. Fading in and out, unsynced, ever-changing—and with a bit of foam on top. To me, the most interesting sounds happen when several sources interact with each other. So I try to always leave enough space in the frequency spectrum so they can co-exist, become one and then separate again.
>
> *(Headphone Commute, 2019)*

Other musicians devise simple ways to generate complex sonic results by combining basic elements that can interact. The producer Seth Horvitz (aka Rrose) makes an analogy to chemical reactions to explain this way of setting up a living musical system:

> I put a bunch of elements together, and if I hear something emerging out of those combinations of elements, that's where I try and investigate further. Instead of getting an idea and thinking, 'I want to make something that sounds like this' and then figuring out how to make it, I'm putting a bunch of things in a pot and looking for some weird chemical reaction to follow. I'll use that as a guide for what I want to do with the sound rather than deciding where it should go from the beginning.
>
> *(Smith, 2020)*

For Horvitz, complexity grows organically from simple systems. For example, by altering a single oscillator in a subtle way:

> If I'm working with a synthesizer, for example, the sort of thing I do now is turn up an oscillator and listen to it and then do something really simple or subtle to it. I try to avoid building complex systems and instead try and get complex results from simple systems.
>
> *(Smith, 2020)*

Horvitz's complex results from simple systems idea is shared by Autechre. Over a career that has seen their work develop into "an epic oeuvre of radical composition and decomposition" (Walmsley, 2024), the duo have spent decades refining patches in Max, the programming environment which they use as a system for making music. Their Max-based system is built from traditional electronic music elements and devices such as oscillators, sequencers, samplers, drum machines, reverbs, and so on.[2] But Autechre use these components to generate a tremendously varied sonic output.[3] Sean Booth explains the simplicity-complexity connection:

> I'm more interested in very simple processes giving complex results than complex processes to achieve complex results. I think it's more fun to get a lot out of very little. In life a lot of things are like that: a lot things we think of as being complex processes and phenomena are underlyingly extremely simple. That's just an intuition.
>
> *(Booth, 2022b)*

In a 2022 Twitch stream, Booth elaborates on Autechre's musical system by explaining his interest in sidestepping the sound of traditional DAW sequencing, in which single sounds are arranged as events along a linear timeline. Rather than being limited to sounds that strictly repeat along a timeline, Booth is interested in a system that generates unique sonic events that bring a sense of *dynamism* (or in Alexander's terms, alive at many levels at once'ness) to Autechre's music:

> I'm more interested in what things are than where they are, from a sequencing point of view. So the shape of things has become something I'm really interested in. A lot of people are thinking about note position because they're used to thinking in terms of piano roll or DAW sequencing or triggering of discrete events that are self-similar. Whereas I'm much more about generating events that are unique or that are context dependent. The nature of the events is much more interesting to me than the position. That's coming from a tape editing background—because I started to realize that what was missing from a lot of computer music or sequenced music in general—most electronic music—was the dynamism. The fact that all the [DAW-sequenced] notes are the same shape is a little frustrating, once you recognize it.
>
> *(Booth, 2022b)*

Some producers consider their musical ideas as input for the musical system in which they're working. Like Eno and Holden, producer Gábor Lázár allows his ideas to ride on the output of his system, bringing him into unknown musical territories. He describes how a system can combine say, rhythmic sequences and modulations to create a non-predefined output: "I try to find a balance between a rhythmic pattern and filter modulations which together articulate some interesting quality" he says (Ravens, 2020). Other producers happen upon a musical system by combining a few effect devices to make a bespoke, composite effect through which to manipulate sounds, a topic we return to throughout this book. Producer Nino Pedone (aka Shapednoise) describes such a system. Pedone configures Instrument Rack devices in his DAW, Ableton Live. Through these Racks he processes sounds from hardware synthesizers:

> I have lots of different analog machines in my studio, including Eurorack modules and oscillator-based synthesizers, and I like to collect sounds from all of these disparate sources and put them into a system that I've developed in Ableton. That involves making my own instrument racks and inputs and processing sound sources through

> them to develop textures that can be recognized as my signature sound. I'll also use software to create various soundscapes, but I resample everything in audio because I like to have the freedom to edit, chop and automate sounds wherever they come from.
>
> *(Turner, 2023)*

Resourcefully using components and parameters

Another feature of musical systems is that they offer the producer an array of components, controls, and parameters that can be resourcefully used. Producer Astrid Sonne refers to this idea when she explains that rather than fitting sounds to pre-existing production goals, "I try to bring out different musical parameters from the material I'm already working with" (Fischer, 2021). Even a single effect or instrument can be a kind of musical system. For example, a reverb device typically has numerous parameters that one can automate to interact with one another, alive at many levels at once. Recognizing this, producer Lindsay Wong (aka Polygonia) makes "full use of the parameters within a single effect" to build her sounds. Wong explains how this technique creates an organic sound from a simple sine tone:

> Even with simple elements, like sine tones, you can create complex textures if you modulate them cleverly. You don't need to stack multiple oscillators—just one, with a well-designed envelope and maybe a delay effect, can generate a rich, organic sound. If you know what each parameter does, you can surprise people with the complexity you achieve.
>
> *(Joyce, 2024)*

Along these lines are examples of producers who use a single instrument as the basis for an entire recording. For example, Andreas Tilliander's 2013 album *TM404* is a resourceful assemblage of vintage Roland hardware. Tilliander used the TR-202 and 303 sequencers, and the TR-606, 707, and 808 drum machines to build a musical system. The track "202/303/303/303/808" exemplifies this system's constantly fluctuating rhythmic syncopations, limited soundset of tuned percussion sounds, and harmonic stasis. *TM404*'s tracks repeat in a way that maintains intrigue.

Or consider producer Gerald Ryder Simpson (aka A Guy Called Gerald), who devised a system to produce his 1988 acid house classic, "Voodoo Ray". Rather than use pre-made drum loops, Simpson synthesized and programmed his own percussion sounds. The main instrument in Simpson's musical system was a TR-808 drum machine. Simpson recalls using the 808 as both the track's sound source and its sequencer, processing each of its sounds with different effects:

> Everything was really organic. There were no turntables involved, there was no DJ'ing involved. It was all synthesis and drum machines and no loops. I'm not saying loops are a bad thing. But I think what's kind of happened now is that people are forgetting the actual skills of synthesis and programming, by getting really lazy and just grabbing [samples].
>
> *(Doyle, 2015)*

Like Wong and Tilliander making full use of their production gear, Simpson was resourceful in thinking about the 808's sound possibilities. He understood that he could build a musical system by combining the instrument with a few effect devices. Such was Simpson's interest in figuring out ways to sonically treat each of the 808's drum sounds with effects processing that he diagrammed his system's ideal processing options:

> You had one snare on the drum machine but you could make that snare sound like nearly anything you wanted. If I wanted a lower snare, I'd pitch the tape machine up a couple of ips [inches per second] and then when I'd put it back down to the normal speed it was like another snare. Then you'd put a reverb on it or something. So there were all these things that you could do. I used to love spending time treating every single instrument. At one point, *I created a diagram of a perfect system,* which was a grid-like snare, bass drum, hi–hat on one side and then on the other side I would have reverb, flange, delay, compression and then I would [connect] one to the other.
>
> *(Doyle, 2015, italics added)*

Almost forty years after Simpson's experiments, processing every sound in one's tracks is a routine production move. Yet even with the DAW's sound-sculpting power, producers seek bespoke musical systems that are both generative and performative. For example, producer Max Cooper considers Sequential's Prophet-6 analog synthesizer a key contributor to his workflow. Cooper uses the phrase *parameter space* to describe the Prophet's timbral-shaping options. He evocatively characterizes the instrument as a musical system whose parameters equate to a multidimensional space:

> The Prophet-6 is my most used synth. It's just got a nice balance of not trying to do too much, but doing enough to be flexible. The parameter space it covers is high proportion beautiful. Any synth has a bunch of parameters with continuous values. If you've got 50 parameters, you've got a 50-dimensional space, basically. That contains all the sounds a synth can make, all the types of timbre.
>
> For me, the Prophet-6 has the largest proportion of beautiful parameter space. There are synths that can do just as beautiful things—and a lot more—but maybe there are parts of that parameter space that I don't really like the sound of and I have to tune them more carefully. The Prophet-6 is just this lovely, crafted thing that has a lot of beauty in it.
>
> *(Truss, 2024)*

Cooper explains the appeal of the Prophet and other synthesizers in terms that evoke both Alexander's alive-at-many-levels-at-once beach scene and Stafford Beer's surfing-esque image of riding on the dynamics of one's system:

> That's what I'm looking for with a lot of this stuff. 'How can I set up a system that has a life of its own?' I can play it and I can still play my chords and express what I want harmonically, because that's where this emotion is, but then it takes on a life of its own and wanders around in these weird ways. I'm always looking for a system that's partly generative and partly performative.
>
> *(Truss, 2024)*

Techniques for setting up a living musical system

With the above examples in mind, if we want our music to sound alive at many levels at once, we need to set up its component elements as a living musical system so that they interact with and influence one another. This will help us create a visceral and panoramic sonic world. But creating a living system is not easy to do—crafting a sense of aliveness is about more than layering in a few sparkly sounds. Skilled producers design music as a living system whose sounds are intrinsically connected, a system in which this influences that influences this, so that each element in a track gives the impression that it's listening and responding the way a musician does. In short, we want our tracks to have a sense of unpredictability. In music, as in life, unpredictability is "the source of all the extraordinary capacities for unbridled creativity in complex systems" (Theise, 2023: 40).

Our source of inspiration is the natural world. Not surprisingly, musicians are attuned to the similarities between nature's complexities and the possibilities of production. Producer Alexandra Zakharenko (aka Perila) finds in nature two metaphors that are helpful for setting up a living musical system: growing and movement. In fact, Zakharenko's idea of production aptly sums up this entire chapter. "I think a lot in terms of gardens and growing sounds" she says. "Seeding sounds in space and letting them grow and evaporate, and letting the traces combine somehow" (McClean, 2023). Zakharenko also records motion sounds in nature—such as cave echoes, wind blowing, and water flowing—to use in her productions:

> I really like to bring those natural movements back into a sound: sound is movement, after all, but natural sounds move differently from something programmed very precisely. It's all about that natural flow, that contradiction or juxtaposition to our city lives. By processing that, I'm constantly trying to heal myself and share it as something very accessible to everyone.
>
> *(McClean, 2023)*

Producer Jacob Collier speaks of his favorite musics as having a sense of depth, texture, intricacy, and tension that evokes the natural world:

> I listen to something and think, 'there's more dimension in here than I can capture in one go through.' So when I create music I enjoy to hide all sorts of elements in and then to make sure that each thing has its purpose.
>
> I look a lot at the natural world. Study something like a flower: this is something that's a complex organism—there's so much depth to it, so much texture, so much intention. Each part of it is designed to capture, absorb, and pull from the world in a different way. But it's also so simple. I think the best music is like that too...That to me is the kind of North Star: really intricate worlds, a hybrid of different sounds, tensions, flavors, sculpts.
>
> *(Tape Notes, 2024c)*

Producer Nils Frahm echoes Zakharenko's and Collier's natural world analogies. Frahm makes music mostly *outside* the DAW—playing piano and classic synthesizers like the Roland Juno-106, working vintage mixing boards, and manipulating temperamental effects like Roland's RE-200 Space Echo.[4] In concert Frahm evokes a Willy Wonka as dub remixer figure in a Steampunk setting, moving among keyboards and effects to manually generate almost every sound through hands-on interaction. Frahm's performing and producing are informed by a commitment to the enchanting instabilities and interactions of an analog gear-based musical system. This system helps him create "billowing" conditions that *foster* sonic complexities, rather than requiring him to *program* them in. In other words, Frahm has devised a living musical system with which to compose. He talks about this system in a way that, again, uncannily evokes Alexander's alive-at-many-levels-at-once concept:

> Since a lot of my gear is analog and all elements correspond with each other, all of the effects I want to apply to the music can quickly be implemented manually. Every piece of equipment influences every other one in some form. And that's something I could never get done with plug-ins. If I put a digital delay on a looped, four-bar beat, the delay will simply repeat itself. But when I send the same loop through a billowing tape-delay, it will sound slightly different on each cycle.
>
> So, in a way, this set-up is creating complexity without me having to force it. To me, this is just the way nature works: a wave on the beach, the symmetry of a tree, the symmetry of your face. The studio sounds different each day.
>
> *(Fischer, 2022)*

In sum, these are but a few of the ways producers create living musical systems. Now let's go deeper to explore techniques that can help us find our own way: live performance of parts, using automation, and modulation processes.

Live performance

Perhaps the simplest way to create a living musical system is to play every note of every part in your tracks. Yes, this is an overly idealistic (not to mention impractical) idea of a production workflow. But it's very much possible to pull off if you overdub parts, one at a time. Playing every note of every part is time-consuming and imperfect, but therein lies its power. The cumulative process rewards the producer with a sound that is the closest approximation to live performance.

Some element of live performance is an essential part of many electronic musicians' workflows. Producer Jon Hopkins uses improvisation as broad strokes to quickly capture the essence of a track, which he refines later: "I work in broad strokes at first—getting the essence of the music down as quickly as possible and leaving things free until I'm happy with that" (Hopkins, 2021). For Hopkins, improvising is a way to *capture reality in the computer*—whether by playing piano or drumming a rhythm on a tabletop and recording the results. Explaining his method for making drum grooves, Hopkins explains how he might "drum on something, as free as possible, and then get it on the screen. It's about getting the humanity into it, breathing real stuff into it. Having one thing gridded and everything else is completely loose" (Turner, 2013).

Similar to Hopkins' getting the humanity into a drum groove, producer and cellist Lucy Railton adds effects to her music by performing them. Rather than using automation, Railton maps effect parameters to a hardware controller which she uses to play reverbs in real time:

> I don't like using automation, for example, in production, like using a keyboard mouse to draw 'organic' lines. I think because I'm an instrumentalist, first and foremost, I feel a lot more connected to manual and physical ways of making music. So in the studio, if I'm adding space to my recordings with reverbs, for example, I would normally record this live using a large hardware controller…This way, I can react in real-time to what I'm hearing, opening the sound to larger or smaller spaces as I feel it. The sound then kind of bounces around in different corners of space. I can throw it where I like; it can feel like you're running up to the sound or it's running away from you. I feel I can only get these results when my hands are in control of the process. It has a bit of an expressive painting effect; the sound is a gesture that I spit out on the canvas. Having hardware tools outside the DAW is essential for this; it's more choreographic, and I feel I can get more visceral results this way.
>
> *(Lichterwaldt, 2024)*

Railton's approach is a brilliant way to set up a living musical system. It's also accessible to any producer, given how easy it is to map a DAW's parameters to MIDI controllers. Moreover, Railton's terminology for describing her tactile philosophy—*manual, physical, hands in control of the process, sound as gesture, choreographic, visceral*—reminds us of the benefits of transmitting one's physicality into a production workflow.

A sequence that never repeats

The easiest way to build a sequence is to record a few notes or measures and then copy them so that the sequence repeats as a block of sound. But in music production, easier is not necessarily better. The problem with a sequence built from a few measures copied to infinity is that it perfectly repeats and doesn't vary itself. Without variation though, a musical part tends to lack depth and mystery as to what it's doing, where it's going, and its capacity for change. A sequence that perfectly repeats is predictable; we already know where it's going because our ears subliminally catch onto its inhumanly perfect repetition. A producer can, of course, disguise sequences so that their sound changes on a surface level (i.e. timbrally by using a slowly-opening filter). This helps, but it doesn't fix the deeper stasis that becomes monotonous. The problem is non-human powered, synthetic repetition. What a producer's music benefits from then, is a sense of unpredictability and the urgency of uncertainty.

To address this, try recording a MIDI sequence by playing it though, from a beginning to an end. When we play a sequence, we solve the problem of synthetic repetition by trying to replicate it and subtly failing. Perhaps we try to repeat a good section—three chords, say—but get one of the chords wrong, or become sidetracked by a variation that's emerged thanks to the misplayed chord. Our playing circles around a musical idea without ever perfectly articulating it, or tries to find an ideal version of an idea without succeeding. But, as producer Andy Turner of the production duo Plaid notes, "often the originality is in the mistakes" (Gillham, 2019). In sum, play your parts to almost find your best themes and their forms to produce a very long sequence that never repeats.

Having eight pairs of hands on the dial: automation

Perhaps the most comprehensive way to create a living musical system is to carefully automate the parameters of the sounds and processing devices in your track. *Automation* defines the value of a control or parameter over time. As a musician vividly notes on the TechnoProduction subreddit, automation is what allows a producer to "touch twist and grab your music as a whole" (TossThisItem, 2022). A simple automation example is controlling a track's volume so it gradually changes over a specified period of time in a calibrated way. But there's no end to automation in a DAW; in fact, most parameters, values, controls, knobs, selectors, and sliders within the software can be automated so that they move by themselves over time (Figure 2.2). Producer Guy Sigsworth sings the praises of automating VST synthesizers and effects, emphasizing the new sound worlds that are possible when a producer changes *multiple* parameters simultaneously:

> I know the anoraks will say that virtual analogue synths aren't as good as the originals, that they don't go out of tune as wonderfully as the originals, but the automation possibilities of effectively having eight pairs of hands on every oscillator, and moving them in a coordinated way, are radically new. You could never have done that on a Minimoog in

FIGURE 2.2 MIDI data like marionettes (photo by Frederick Stubbs)

> 1973. I think the automation of plugins, reverbs and things like that, is a fantastic possibility, because it gives us a chance to manipulate musical space in unprecedented ways. Again, it's like having eight pairs of hands on the dial of a Lexicon [reverb] or something.
>
> *(Inglis, 2001)*

What do we automate? *Everything*. If a parameter can be automated, experiment with doing so to hear whether or not its movement creates enchanting, alive-at-many-levels-at-once effects.

Modulation

Modulation is a subset of automation that uses a modulation *source* to change a parameter's value over time. In electronic music production's pre-digital era, musicians used components of analog synthesizers such as LFOs (low frequency oscillators) and envelope generators to modulate various parameters of the instruments. An LFO, for example, generates an *ultrasonic* oscillation at a very low frequency beyond the range of human hearing (Strange, 1983: 12). Musicians have long relied on this oscillation device to add movement and pulse to sounds by modulating parameters of pitch, amplitude, and cutoff frequency. When used to modulate a sound's cutoff frequency, for example, an LFO creates a pulsating, *wuh-wuh-wuh* effect. But there are many other ways of using LFO modulation. Producer John Hayes explains how he experiments with assigning LFOs to various destinations within Ableton Live:

> Ableton's LFOs are all over the place when I am working. I love their ability to bring an atmosphere or a space to life by adding some movement and variation. I'll have dedicated sessions where I'll just see what I can assign them to and see what kind of interesting sounds I can create.
>
> *(Headphone Commute, 2023a)*

Drew Daniel of the production duo Matmos stacks LFOs. Daniel uses a Max for Live device by Kentaro Suzuki called *LFO-Cluster* to create layers of LFOs that modulate multiple parameters in his Ableton samplers:

> It's a solar system of cascading LFOs that are related but distinct to each other, and you can map them to parameters. I'll fill up Sampler or Simpler in Ableton, put some EQs and filters on, and then flood every single parameter with different LFOs chained to other LFOs using the LFO-Cluster technique.
>
> *(Turner, 2024)*

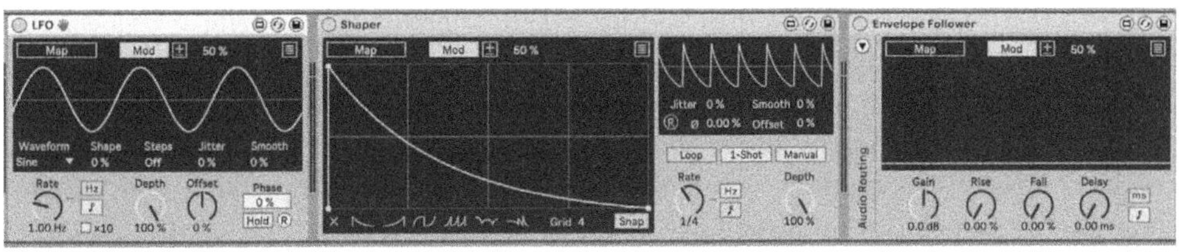

FIGURE 2.3 Ableton Live's LFO, Shaper, and Envelope Follower devices

Producers use modulation techniques to animate their sounds and bring a sense of life and unpredictability to their productions. This is further helped by the fact that in most DAWs modulation devices are multi-mappable, which means they can be *mapped to control multiple parameters simultaneously* to create what one article on the topic calls "captivating sonic interactions—a living, breathing sum greater than any individual sound" (Noah, 2024). Ableton, for example, offers *LFO, Shaper*, and *Envelope Follower* modulation devices (Figure 2.3), all of which are multi-mappable. Shaper is a tool for creating multi-breakpoint envelopes that generate data to modulate musical expression in a mapped parameter. And Envelope Follower analyzes the volume pattern of an incoming audio signal's transients and then outputs this pattern as a continuously changing control signal that can be used to modulate other parameters in the DAW. A producer who wants to create even more musical possibilities can use the "random" functions of such modulation devices, or connect multiple devices so that modulators modulate one another. One could, for example, control the *rate* of an LFO with the *contour* of a randomized envelope tool like Shaper. This is what Daniel means when he evocatively speaks of *a solar system of cascading LFOs.*

Daniel's cascading LFOs idea is echoed in a 2023 discussion on Elektronauts, an online forum run by Elektron, an esteemed electronic musical instrument company. A producer posts a link to a 2017 track by Calum MacRae (aka Lanark Artefax), "Touch Absence", and asks: "How I can go about achieving this sort of sound?" Having studied the music and noticing in it "a huge variety of sounds that are organic and dynamic", "less static instrumentation", and "big developing sounds", the producer is both inspired and "curious as to what [techniques] exactly gets you here" (dokev, 2023). It's an excellent question. An interesting answer comes from a fellow producer who suggests incorporating multiple levels of modulation into one's productions:

> Fundamentally, if you want less static instrumentation and big developing sounds, *you want to be modulating your stuff, and modulating the stuff modulating your stuff.* Having the focus be on a very few elements, but having those few elements constantly morphing over time through modulation etc. seems to be the recipe here.
>
> *(dokev, 2023)*

Because of the way modulation devices are integrated into DAWs, techniques of modulation, while not new, have become sonic hallmarks of producers who imaginatively configure their music as a living system. For Ian Dicke, a composer who builds Max for Live plug-in devices, modulation is an alternative to the endlessness of automation's "unprecedented control over every parameter" in a DAW and a way to "break the shackles of constant knob tweaking" to relinquish some creative control in the production process (Noah, 2024). Another producer and Max for Live developer, Alessandro Miracapillo (aka Aylesim), understands modulation as a way of making surprising artistic choices one may not have otherwise made. Miracapillo's *Knob Studio* device aims to "bridge the gap between my initial ideas and unexpected outcomes" (Noah, 2024) And producer Thessa Tossing (aka upsammy) sums up modulation's general appeal for producers who want to make music alive at many levels at once. By fusing randomization and modulation in the DAW, she says, "you can really create as if [the track's] an organism, it can almost write itself" (Davis, 2024).

Finally, for producer Sam Shepherd (aka Floating Points) modulation is a way to bring life to synthesizers that are otherwise a step removed from their performers. Unlike an acoustic musician who has developed a physical connection to a physical instrument, a topic we'll explore in Chapter 5, producers can feel disconnected from their synthesizers—as if the instruments are partially inaccessible to them, hidden behind hardware circuits or software code. Shepherd's production goal is to use techniques of modulation to bring the "soul" of his synthesizer parts to life and thereby connect the performer with the listener:

> When I listen to musicians—when I see a cellist embodying a cello—I want to hear into their soul. I feel that's fairly easy, however good you are as a cellist, because you're connected to that object. But with a synthesizer it's a lot harder because you're behind all these circuits. So I long for electronic music where you hear beyond the instrument into the soul of the performer, and my entire search, my entire practice is *that*. So sure, I'll dial in this sound, but you have to play with the modulation whilst you're playing it to convey that soul of the instrument.
>
> *(Tape Notes, 2024d)*

Music's main enchanter: hearing human touch

> I always think that synthesizers have no business being touch sensitive, actually. A piano is a piano and the touch sensitivity on a synthesizer is always going to be artificial. It's the physical thing of a piano and it's going to be somebody's interpretation of it on a synthesizer—and different on every synthesizer—which may not suit you. My [PPG] Wave and my Minimoog have no touch sensitivity and it's almost like they're more real that way. Synthesizers are great but it's no good using them as pianos—they're a thing apart, different.
>
> *(Anne Dudley, founding member of Art of Noise)*[5]

When Drew Daniel talks about flooding *every single parameter with different LFOs* he's answering the question of how to make music alive at many levels at once. Another angle on this is that we can understand production practices as a means of imbuing our music with a human touch. If you produce using DAW as your primary instrument, as many musicians do, a sense of touch is lost in the gap between the sounds we *intend* to make and the sounds we're *able* to make. There's a disconnect between doings and soundings: our sense of musical touch feels blunted, and our instinct for how to proceed—which in a non-computer context might lead us to play something on an instrument and build on that—is stymied. What should we do?

Learning how other producers conceptualize their workflows foregrounds this disconnect between doings and soundings. One production principle musicians talk about, albeit obliquely, is the importance of *making music that captures a recognizably human touch*. When musicians talk about making their music more *organic*, more *alive* and *breathing* and less robotic and cold, they're talking about allowing their humanity to be heard in their sound. In addition to modulation techniques, there are other ways to achieve this human touch, or a semblance thereof. Let's consider four approaches: using samples, thoughtful repetition, evoking the analog sound of production's earlier eras, and by recording oneself playing an instrument.

Using samples

Using samples is a time-tested way of teleporting human touch from other times and places into our musical present. Whether you're using a kick drum from an old record or a bit of subway ambiance recorded on your phone, sampling, observes producer Christina Vantzou, sends "information from a specific place, maybe together with a memory and/or a moment of intense imagination" and imbues the music with this touch (Rose, 2023). For producer Amon Tobin, sampling is a practice of capturing the energy of other recordings, akin to how a sports photograph freezes an athlete in motion. Even tiny "particles" of recordings, Tobin explains, are microcosms that contain the energy of their source:

> If you look at sports photographs—someone in mid-air jumping. You can tell what happened before and what's going to happen after, but in that frozen moment you have all the energy of both things encapsulated. And that's more or less how I viewed sampling: it managed to trap the energy of something much bigger than its little components. Then when you re-contextualize that and you put it amongst lots of other things that are pulling in lots of other directions you end up with a really dynamic and interesting sound. And I found that you can do that with smaller and smaller and smaller particles and they'll still retain some energy of something before.
>
> *(My Forever Studio, 2023)*

Producer Geir Jenssen (aka Biosphere) says that the sampler is his most important production tool because it allows him to use samples of "organic" acoustic instruments as material for sound design:

> Many of my sounds are based on short samples of classical, acoustic instruments. Preferably from the quiet parts of symphonies where the musicians play carefully and where you can often hear the chairs creaking. When I amplify these sounds and run them through various filters and envelopes, I often get sounds that are almost impossible to make on a synthesizer. So the sampler has been my most important tool all along. I often start a composition by creating a theme with an acoustic-digital filtered sound. Over this, I often try to integrate pure synthetic sounds from a synthesizer. This makes the music sound a little more organic than if I just used synthesizers."
>
> *(Headphone Commute, 2021a)*

Building on the ideas of Vantzou, Tobin, and Jenssen, every producer can capture the imagination, energy, and human touch of their own work by making samples of it (i.e. resampling) and re-incorporating those recordings back into tracks.

Thoughtful repetition

The reader may rightly be wondering, What is *unthoughtful* repetition? Like the cut and paste sequence we explored earlier, it could be a part that never changes, varies, or evolves and so, while it may be functional, it's lifeless. To counteract this problem of synthetic repetition, producers who work with beats go to great lengths to capture and finesse a sense of the human touch through subtleties of performance and programming.

Pioneering producer Jean-Michel Jarre, who famously sellotaped the buttons on his Korg Mini Pops drum machine so that it played two different rhythm presets at the same time (Simpson, 2018), explains how he eschewed strict repetition in his drum loops and sequences. Jarre wanted to avoid what he perceived to be a "robotic sound" in the music of some of his contemporaries:

> When I started to do electronic music, I was obsessed...about not having anything being repeated in exactly the same way. For me it was exactly the opposite attitude to that of Kraftwerk, Tangerine Dream and all those electronic bands who were doing something more robotic. I considered electronic music in a much more sensual, organic way, where nothing should be repeated.
>
> *(Flint, 2008)*

Escaping a robotic sound is standard practice today, thanks mostly to the rhythmic finessing practices innovated by hip hop producers since the late 1980s. Even though instant note correction or *quantization* (i.e. rounding off note locations to a prescribed note value, such as a 16th) is available with a keystroke, beat makers routinely experiment in the DAW with various kinds of groove feels ("Straight" or "Swing"), in varying amounts (100 percent or 20), and applied in different contexts (kick and snare, or only the hi hats). Alternately, producers *play* their touch into tracks. For example, legendary producer James Yancey (aka J Dilla) was renowned for combining unquantized free-hand finger drumming with various quantization values to achieve unique groove feels (cf. Scarth and Linn, 2013; Charnas, 2022).

Evoking analog sound

Another way to conjure the human is to use analog circuit-based electronic instruments or effects processing that reproduce the sound artifacts associated with vintage gear. Examples of this include analog synthesizers and drum machines, plug-ins that faithfully emulate the wow-flutter wobbles of magnetic tape recorders, mixing consoles with their quirky, no-two-are-alike channel strips, classic valve and tube compressors, plate reverbs, crappy RadioShack speakers, and so on. Today, every part of yesteryear's recording studios is emulated.

How does old gear produce sounds considered more human-touched than sounds forged digitally? One answer is that the electronic signal paths of analog equipment and their components introduce *artifacts* into the sounds passing through them in the form of micro-saturations and harmonic distortions. It's the imperfections and inconsistencies of these artifacts that have come to be perceived as sounding warm, and therefore un-machine-like. Put another way, since the late 1950s the sound of vintage analog gear's warm mediations have been heard as faithfully reproducing real life. In other words, the analog sound has become naturalized. Production educator Eddie Bazil explains that while "the [digital] signal clarity we have today can come across as clinically staid...we actually miss the imperfections of yesteryear's sound" (SOS Recording & Mixing, 2025). Similarly, for producer Cristobal Tapia de Veer, the saturation and compression of analog gear creates a coveted "fat" sound that "is just more natural to the way we hear things" (D'Silva, 2023). And so it's interesting how in recent years the sepia aesthetics of imperfect-sounding vintage equipment and their simulations that conjure a human touch have become integral to the sound of many varieties of lo-fi music. For example, in *lo-fi beats* and other ASMR easy listening/chill styles producers foreground imperfections of performance, timbre, and recording resolution to conjure a cozy sound—an aesthetic of intimacy. In sum, the producer has many ways of incorporating an analog/lo-fi sound into any aspect of a track.

Recording oneself playing an instrument

A way to incorporate your touch into your work is to be deliberate about how you compose a track so as to preserve and amplify whatever touch you have. In *The Unknown Craftsman*, a classic book about the value of handiwork, art critic and philosopher Sōetsu Yanagi reminds us that while machinery "gives speed, power, complete uniformity, and precision", it lacks our hands' sense of touch. "No machine can compare with a man's hands" he says. "It cannot give creativity, adaptability, freedom, heterogeneity. Man prefers the creative and the free to the fixed and standardized" (Yanagi, 2013: 108). For Yanagi, craft is all about the hands because it's through our hands that our touch imbues our work with a sense of flexibility and difference. *In music we want to hear the hands that made it.* Producer Bogdan Raczynski amplifies this idea by reminding us that the musician is more important than the musical machinery. "I am the technique. I am the instrument" he says, and rightly so.[6]

We hear touch in music. Think about touch as a conveyor of your body thinking in the music as you're making it. This may or may not rule out using conventional loops and sequences. Just remember that prefab musical structures leave less room for human maneuvering as their perfect synced'ness has made them impermeable to error.

But we want error! Listeners want to hear struggle or striving inside a music's process because it's struggle and striving that bear traces of the producer's touch. Recording improvisations is a potent way to engage struggle because when we improvise, we have to think in the moment, quickly and imperfectly. When we improvise there's an audible sense of urgency, a sense of something at stake.

So: try recording a part for a track freeform, *without a click track*, and then fit other parts around the performance's odd timings. This fitting together will take some craft—maybe you'll make the other parts repeat, but not quite loop, to complete the arrangement jigsaw. This way of working captures and builds upon the unsettled energy of your track's initial steps. Producer Loraine James incorporates this workflow, free improvising first and condensing it later. "Sometimes I don't like thinking, *this part has to be exactly four bars*, and so forth" she says. "Sometimes I'll hit the record button and just go, ignoring the metronome and just doing whatever. I'll improvise for five minutes and condense that into a song" (Mullen, 2022).

Producer Mark Brandon (aka Model Man) also records beats and long takes of ambient keyboard playing, without using a click track:

> I've done a lot of stuff where I've played live drums without a metronome and then worked around the drums. Also I do a lot of ambient stuff that has no grid. So I'll just play. I've got a Korg M1 that I'll just play and record for an hour and take things from that. I think there's definitely something to be said for ignoring the grid from time to time.
>
> *(Mullen, 2023c)*

Improvisation is just one of many human touch markers producers can incorporate into their work. Other markers we might try to capture have to do with uncertainty and failure:

when a part almost arrives somewhere, but not quite;
when we can hear the influences of another musician in the sound, but the influence is warped;[7]
when we repeat the good bit twice, but play it wrong the third time because we couldn't pull it off;
when don't know how the music will end up and incorporate that uncertainty into the music's form.

In other words, find ways to turn your technical liabilities into assets.

In sum, producers do well to *prioritize ways of nurturing and articulating what for them is the human core of their music.* We want to hear this core in the music—it's why we listen. Touch is music's main enchanter and the reason why music's affective metaphors are always physical: we listen to be *moved* by music, *stirred* by music, *touched* by music.

Populating sounds 1: working with presets

A place to begin configuring a living system that generates beautiful outputs is to create variations on a single sound so you have multiple sounds to work with. Start by diving straight into the abundance discussed in Chapter 1, browsing the presets in your DAW and VST instruments, not looking for anything in particular, just waiting to be surprised. Sounds are everywhere in and around the DAW and most virtual instruments include hundreds of presets that demonstrate their range of timbres. Presets situate music software at the intersection of the possibilities of your practice and the possibilities of your musical system.

Some producers are weary of presets, believing that using them is a creative shortcut, or worse, evidence of one's lack of sound design skills. In his book *Presets*, Stefan Goldmann explains the "unwritten rule that creativity in electronic music means, to a greater of lesser extent, engaging with sound design" (Goldmann, 2015: 12–13). Producer Nathan Micay speaks for many when he admits, "When I first started, I used to think sampling was cheating, and I used to think presets were cheating" (Mullen, 2023a). But Micay had a change of perspective when he began identifying presets in the music of other musicians. For example, referring to the free sample packs of Danny Wolfers (aka Legowelt), Micay realized that "once [Legowelt] started to give out these sample packs of presets from those synths…I heard all those samples in Four Tet songs. I was like, It's not cheating at all! Nobody cares" (Mullen, 2023a). Indeed, few musicians listen to a track and think, *Is that a Legowelt kick drum sample?*

What producers *do* care about is having ways to quickly choose and dial in sounds for their tracks to keep a workflow moving. Producer Mike Paradinas (aka µ-Ziq) sees presets as solving "half the battle" of building a track. For Paradinas, sound designing in the sense of creating original sounds for each part in a track is *not* essential to music production:

> I am not one of those producers that considers themselves a 'sound designer'. I hate the term, actually. It makes me think of cinematic rumblings, car park door slams, and Hans Zimmer—all that does nothing for me. I am a musician. I like to use what's at hand to create music. Using a preset in a more creative way can do more to advance music scenes sometimes.
>
> *(Headphone Commute, 2023)*

If you work with software instruments, it's worthwhile to get to know their sounds, and the best way to do this is to search through their presets to learn which ones speak to you and therefore could be useful in your work. "If I like a preset" says producer Sibel Koçer (aka JakoJako), "I try to change it to my needs" (Fischer, 2023a). Following Koçer, listen to a sound, and if you like it change it a bit, rename it, and save it so you can locate it again. As Tiny Johnson (aka Shima) reminds us, "Even if you're using a stock preset, but you tweak it just a little bit, save it. That's how you start to build your own sound and that's how you become recognizable" (Lee, 2023). Searching for sounds can lead us to discover materials which may one day be used in a track.

Of course, many presets are more miss than hit because most presets don't suit our tastes. Their sounds may be too obvious or one-dimensional. In the 1980s, the film composer Hans Zimmer explained that he doesn't use presets because synthesizers lack the responsive variability of acoustic instruments played by live musicians. Unlike a musician who "will change the sound drastically from one track to another", a preset is static—it doesn't respond to its environment. A preset, says Zimmer, "will never be as perfect as if you tweaked it and actually spent a bit of time on it" (Denyer, 1986). Alternately, a preset's sound may be so *multi*-dimensional that it hogs mix space and is therefore difficult to blend with other parts in our music.[8] A preset comprised of multiple oscillators and layers of modulated effects, while impressive, lacks ambiguity. There's not enough *missing* from the sound to spur our ears to fill in the difference. Still, we search.

But even presets that miss more than hit can help us understand what sounds we actually like. You might like a sound, except for some of its details which could be improved. It's common for the presets in a virtual instrument to be heavily processed because this shows off the software's capabilities. But the overblown is a hindrance to making music. Producer Jon Hopkins describes how he'll *remove* effects from a preset to get it down to its core element and learn how it works:

> Sometimes it's nice to just start something from scratch. I will pick a preset and usually—because everything has got so many layers in that synth—I'll immediately turn off the effects. I'll immediately turn off two of the oscillators and just hear just one because I need to know what's going on...I will usually strip away loads of things and find one element that I love.
>
> *(Hanging out with Audiophiles, 2023)*

Following Hopkins, re-shape a preset by experimenting with reducing the number of effects, or playing with other parameters you notice on an instrument. A simple sound re-shaping technique I use often is *remove the reverb*. This is sound design! The principle here is that it's by removing, not adding from a preset that we reveal sonic worlds. As Nassim Taleb reminds us, "knowledge is subtractive, not additive—what we subtract (reduction by what does not work, what not to do), not what we add (what to do)" (Taleb, 2010: 78).

Populating sounds 2: sound tweaking

When you encounter a preset that inspires you to want to work with it, start tweaking it. To tweak is to make very simple alterations to a sound—it's sound design writ small. One might change its ADSR (attack decay sustain release) parameters, open up a filter, add a whispering of delay, or narrow the reverb's stereo width as producer Nathan Fake does. "I love the sound of mono-ed reverb, that's the reverb width set to zero" he says. "It also adds a lush vintage quality to the sound and overall mix" (Mullen, 2021). Try things. If the sound has a pulsation element, find its source (which is often an LFO) and then alter the speed and intensity of that pulsation. It's by trying out small changes that a tweaked preset often becomes less obvious, and therefore more interesting. Once you've tweaked a sound you've customized it to be your own.

To take the process a step further, make additional tweaks to the new sound, and re-name preset variations with consecutive numbers (e.g. pad 1, pad 2...pad 12 etc.) or apt descriptive words (e.g. pad 12 noisy). Re-naming sounds is a simple way to both document and scale up one's sound design work. The music software company U-he, whose sound organization categories we discussed in Chapter 1, encourages the re-naming of presets with its "auto-versioning." When one re-saves a sound on U-he's synthesizers a number is automatically added after the new preset name, anticipating (and encouraging) the producer's variation-making technique.

There's nothing radical about altering software presets to make them your own and saving the results, and some production purists still insist on creating elaborate timbres from scratch—starting with a single oscillator say, and building upon it. But altering presets is an undeniably useful technique for making sense of the DAW's sonic abundance. This practice is both sound designing writ small and a way of engaging with the landscape of sounds in our software as we move towards originality. While we may never know all of these sounds, we can at least get to know a few of them well. Moreover, altering presets expands the producer's collection of useable sounds. Creating multiple variations of a sound *populates* our musical system with meaningfully related materials to work with. Over time the User sections of my virtual instruments have grown into quirky libraries of favorite sounds. A particularly evocative sound or effect in the library is easy to spot because it has anywhere from five to twenty-five variations I've spun from it.

What might a musician learn from building a collection of varied yet related sounds? Here are three things. First, a preset with many variations is a reliable indicator of a sound's interestingness and importance to our work. The fact that we've taken the time to make many variations of a sound suggests that there's something about it we find intriguingly musical. These variations—or future variations thereof—could be useful one day.

Second, as a brief foray into sound design each preset variation is experimentation within reach of producers of all skill levels, a way into micro-mastery of one's production tools (Twigger, 2018). Admittedly, I would like to learn my virtual instruments more deeply, but that may never happen because I prioritize making new music over fully knowing the software I work with. Producer Bryan Müller (aka Skee Mask) explains the value of having a partial understanding of one's production gear. "I just don't really want to understand a device completely" he says, "because I'm afraid it might take away too much of my natural enthusiasm for experimenting" (Telecom Electronic Beats TV, 2017). Making variations by changing one parameter at a time on a preset is small-scale experimentation that helps us incrementally understand what we can do with our musical system.

Finally, making preset variations is *a practice we practice whilst making music*. Even as we've settled on a sound that works (for now), by saving a few more variations for future use we step out of the compositional frame in a metacognitive moment of self-observation. Pause your process to save some variations. This reminds you that as exciting as today's work is, it's also shaping the conditions for tomorrow's session.

Populating sounds 3: sound fishing

> Ideas are like fish. If you want to catch little fish, you can stay in the shallow water. But if you want to catch the big fish, you've got to go deeper. Down deep, the fish are more powerful and more pure. They're huge and abstract. And they're very beautiful.
>
> *(David Lynch,* Catching the Big Fish*)*[9]

At the outset of a session I often wonder:

How long will I spend searching for the instrument or the sound?
Should I finesse the sound into something compelling or just get on with it?

Such questions relate to the producer's relationship with randomness. Sometimes one searches through sounds, stops on the first one that sounds promising, and then begins making music with it. This approach can feel haphazard—as if we haven't spent sufficient time considering what we're going to do. But it generates results. Since we could not have foreseen the promising sound that inspired us to make something with it, we trusted randomness to guide us. As producer Jon Hopkins observes about this trust in workflow: "Once you have the epiphany that you can't imagine all your own sounds, but they'll come through if they're what the music needs, then there's a great freedom in that" (Mackintosh, 2021).

But producers also steer randomness, which brings us to the metaphorical idea of *fishing for sounds*. In this analogy, sounds are fish, the DAW and its virtual instruments are the pond, and the producer needs to stock this pond with sounds ahead of time. The easiest way to do this is to set up a few virtual instruments in one's template, as discussed in Chapter 1. Each of these instruments is itself an environment: down deep in the instruments I use are many sounds I've yet to encounter, and endless options for altering these sounds into new ones. By stocking the sound pond with just a few instruments, a producer configures a sonic world whose dimensions, like Lynch's ideas, are potentially huge and abstract.

As far as fishing in the sound pond, the workflow is to start somewhere—*begin anywhere* as John Cage used to say—by opening an instrument and playing with it. Sometimes a sound suggests ways of working: we play a part to explore how the sound reacts to our playing and in that process our sound fishing becomes composing. In music production, sound and ways of playing and composing are inseparable: *what you play and compose is shaped by how your sounds sound.*

I'm always surprised to hear sounds I made and then forgot about. As with the experience of playing an acoustic instrument, one never gets to the bottom of music software—one never learns all that it can do. And like a musician's sense of touch, a sound is always in progress: once you've made a sound you can tweak it indefinitely, refining it towards ever more sensitivity and uniqueness. Some virtual instruments, such as Spectrasonics' *Omnisphere*, are oceans of possibilities unto themselves, their extensive banks of sounds types like deep water coral reefs, awaiting discovery.

Once you have one sound you like, you can fish for others by opening another instrument, or another instance of what you're already using. This brings us to the idea of *sound compatibility*. Producers often seek something that *goes with* what they already have. This can mean many things. The relationship between the sounds can be euphonious or contrasting. Where an initial sound is thick, a second sound can be thin, reverb'ed instead of dry, and so on. Keeping sound compatibility in mind, producers often seek unusual textural contrasts and composite timbres. A classic way to do this is by *using the first sound's MIDI to trigger a second sound's part.* For producer Ben Lukas Boysen this practice generates a composite timbre with dynamic layers:

> I often start with a very simple idea, then I double the track several times and have different synths play it...I will play with volume and other parameters of layers to make it more organic and evolving. This method helps trigger new ideas and to develop tracks quickly.
>
> *(Ableton, 2017)*

Finally, sound fishing's aim of identifying the most useful sounds at hand helps us understand an instrument's utility in our music. Sometimes an instrument has only one function. For example, it could turn out that you use it primarily for its bass or granular pad sounds and ignore everything else it can do. This is the way of an efficient producer who knows where to find *this* sound for *this* purpose. Make time to practice sound fishing to get to know the useful sounds in the instruments you encounter, ignoring the unuseful ones. Sound fishing helps us discover what we didn't know was there until we hear it and realize, *I want to use this sound!* In sum, instead of trying to control what you'll find before you find it, set up a sound pond, fish in it, and see what comes to the surface.

Populating sounds 4: thinking genetically

In addition to saving preset variations and fishing in the timbral waters of one's virtual instruments, a producer can make use of "genetic" functions in music software that allow for growing new sounds from existing ones. For example, on the Genetics page of Synapse Audio's *Obsession* synthesizer two existing patches can be selected to generate a composite with characteristics of each (Figure 2.4).

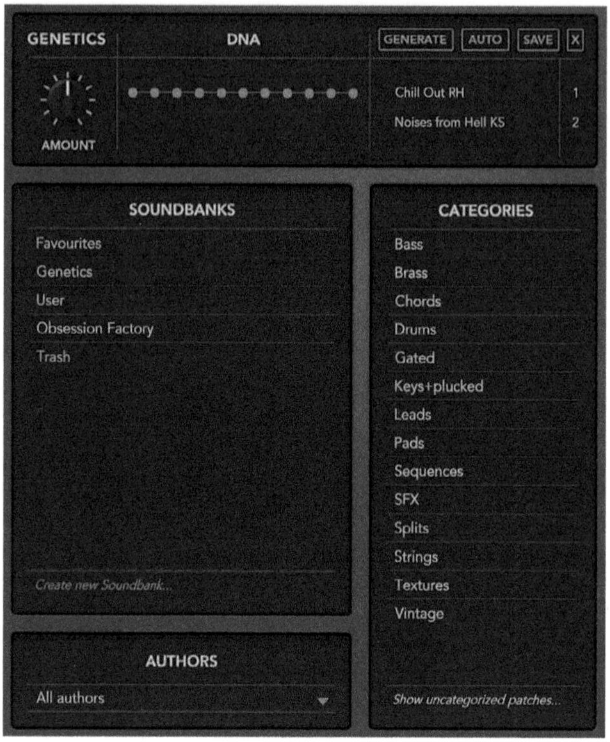

FIGURE 2.4 "Genetics" function in Synapse Audio's Obsession synthesizer

Another example is Synplant 2, a synthesizer developed by Magnus Lidström's company Sonic Charge. Synplant incorporates an elaborate approach to generating synthesized sounds from audio recordings and has two genetics-themed features: DNA Editor and Genopatch. DNA Editor allows the user to "fine-tune the genetic makeup" of a sound. In this window, a synthesized timbre is represented as a DNA spiral with 48 genes, each of which controls parameters pertaining to the sound's envelope, LFO, oscillators, filters, and effects. Genopatch uses AI to analyze source audio and create corresponding synthesized timbres, "by making educated guesses about optimal synth settings, listening to the outcomes, and iteratively refining them to match the reference audio better" (Lidström, 2024). In the Genopatch view, timbre variations based upon the audio source material are represented as vertical strands that "sprout and grow" (Figure 2.5). Unique among the many virtual instruments available today, *Synplant 2* makes sound designing feel like a living production collaborator. As Producer A.G. Cook explains, Synplant can generate random inspiration:

> I am drawn to approaches that are slightly randomized or feel like a kind of collaborator throwing something at me and then I can tweak it and do other things and make decisions. I think that plugin [Synplant 2]—it's just a bit of that little two-minute waiting game when it comes up with things and all the snapshots. It's really fun.
>
> (*Tape Notes, 2024a*)

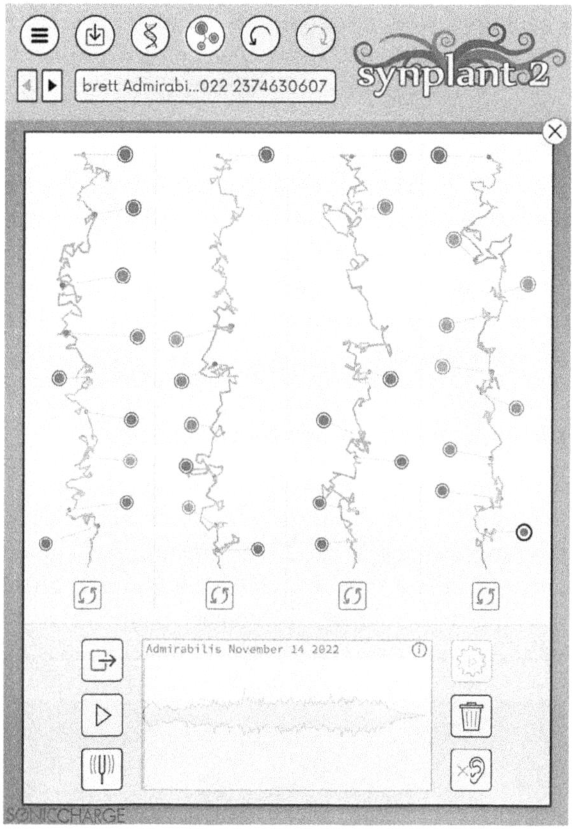

FIGURE 2.5 Genopatch view in Sonic Charge's Synplant 2 synthesizer

In sum, explore the possibilities of genetic variations. Using an instrument's genetic controls encourages us to think of sound design in terms of spinning variations on the sounds already in a track.

Playing an acoustic instrument, playing a DAW

In his trenchant essay *What I've Learned From My Musical Instruments*, computer scientist and musician Jaron Lanier compares playing acoustic instruments to interacting with virtual ones. For Lanier, there's no contest: acoustic instruments always win. The main differentiator has to do with acoustic instruments' magic and mystery. In a word, their *presence*:

> Computer models are made of abstractions—letters, pixels, files—while acoustic instruments are made of material. The wood in an oud or a violin reflects an old forest, the bodies who played it, and many other things, but in an intrinsic, organic way, transcending abstractions. Physicality got a bad rap in the past. It used to be that the physical was contrasted with the spiritual. But now that we have

information technologies, we can see that materiality is mystical. A digital object can be described, while an acoustic one always remains a step beyond us.

(Lanier, 2023)

Playing an acoustic musical instrument that is always *a step beyond us* is a model for making electronic music, and producers who approach their DAW as an instrument seem to intuit this insofar that they develop resourceful and idiosyncratic ways of creatively *playing* the software rather than merely *using* it, finding ways of what Lanier calls *transcending abstractions*. With this in mind, let's compare and contrast the experiences of playing an acoustic instrument and playing a DAW to explore the omnimusical demands DAWs make on musicians.

Playing an acoustic musical instrument: touch and time

Musicians get immediate feedback from their instruments. As a percussionist, my instruments respond to me as I play them: striking creates vibrations, which create sound. Gong playing offers a clear example (Figure 2.6). When I play a gong there's many variables besides the gong itself that shape the instrument's sound. There's the mallet I use (soft or hard, yarn-wrapped or rubber), the contact point where I strike (center, off-center, at the outside edge), the stroke type (single hit or continuous roll), and the playing

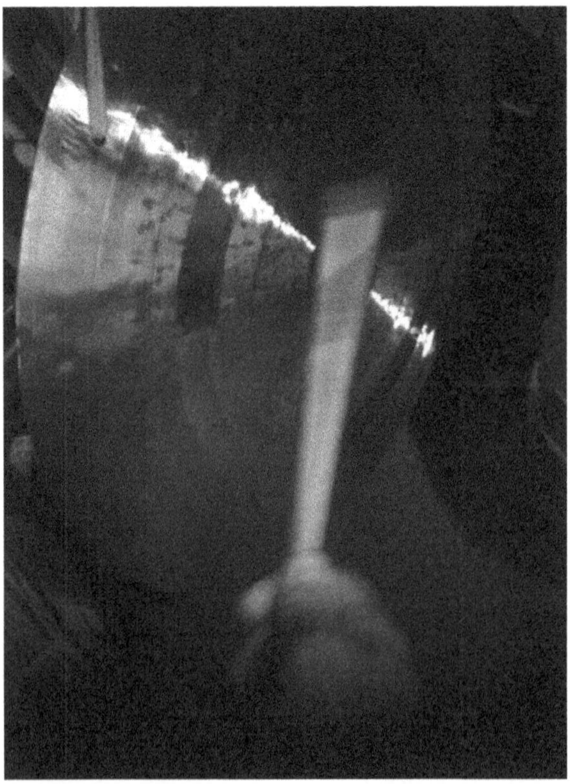

FIGURE 2.6 Playing a gong

intensity (soft or forcefully weighted). Changing any of these variables changes the sound the gong makes. If I want to shift from making a thin and bright sound to a thick and dark sound, I change where and how I strike the instrument. This shifting among sounds is second nature, but it's only possible because of my somatic history with the gong. Such is the close connection musicians feel with their instruments as they play them that their gestures connect to an array of sound moods—or at least, that's the sensation. From the perspective of its player, a gong can conjure a drone, or a distant storm, depending on our sense of touch and our imagination.

The most important aspect of playing an acoustic instrument is the tight coupling between touch and time. Perhaps this is an obvious thing to point out, but at any moment musicians are their own conductors. They can change course on a whim—accelerating or slowing tempo, playing louder or softer, choosing different pitches, finessing different timbral shades via varied techniques—and their instruments respond instantly. Consider another instrumentalist example, a violinist. When you watch a violinist playing a melody, notice when they dig in with the bow and apply more pressure and vibrato to the strings with their fingers to make the melody sing. When we play an acoustic instrument, we're in full control and feel as if one with our gong, with our violin. There's no latency between decision and sound, and there's no hiding—our sound is the sound we have the skills to make. We sound as we are.

Playing a DAW

When I play a DAW and the virtual instruments therein, I have control over sound generating, but the process of making sounds and receiving feedback from my playing is greatly slowed down. As with gong playing, there's many variables shaping the DAW's sound, but these variables are not immediately at hand. Templates, instruments, and sounds have to be set up one by one (and often laboriously, by mouse-clicking, selecting etc.) before I can play them, and effect racks have to be configured before I can hear how they process my tracks. In other words, to get a sound going, a sequence of steps is required.

For example, I might want to work with a pad. I'll begin with a sound that's in the ballpark of what I want, knowing that I can change its sound by altering some of its parameters. Unlike a gong whose fixed materiality mostly determines its sound and provides me with immediate feedback, my pad's sound isn't readymade, nor is it fixed; in a DAW, sounds are always questions open to reformulation. Moreover, finding, designing, and manipulating sound takes time away from making music with the pad! This is why producers sometimes separate sound design work from composing work—because each involves a different mindset. The most important aspect of playing a DAW is not the tight coupling of touch and time, but rather their *uncoupling*.

But there's an upside here. To play a DAW is to trust the process of interacting with the software so that we may accidentally discover something interesting whilst playing with sounds, using effects, say. Producer Scott Morgan (aka Loscil) speaks of "that immediate feedback of what [sound] does when you bend it and shape it" (Hogwood, 2024). And producer Brian Leeds (aka Huerco S.) describes this feedback and interaction as "tinkering away, conducting experiments, & discovering artifacts deep deep below" a track's surface (Arnott, 2013). As we learned in Chapter 1, production discoveries often emerge as a by-product of sounds that have been put into dialogue with one another. The production goal, to quote Brian Eno quoting cyberneticist Stafford Beer again, is to *ride on the dynamics* of a musical system to create music.

This brings us to the questions of what makes a DAW a unique musical instrument, and the unique challenges involved in playing it. A DAW's uniqueness comes from that fact that, unlike a gong or a violin, it offers more than a one-to-one, *this-gesture-creates-that-sound* relationship. Instead, a DAW presents the producer a *one-gesture-can-create-many-sounds* relationship. A DAW isn't a single, fixed instrument, it's a thousand unbounded instruments—an instrument whose instrumentality is always in flux. The DAW exemplifies how "our available sonic palette [has] expanded so broadly as to be functionally infinite" (Sherrill, 2025). What this means is that to play a DAW is to trust the process of accidentally discovering something interesting. To play the DAW is to craft musical experiences both in the moment and over longer spans of time, from live performance to long arrangements, from micro edits to macro organizing, from recording one part right now to shaping many parts into a track over weeks or months of cumulative, touch-decoupled-from-time work.

In sum, playing an acoustic instrument as a model for how to approach a DAW only takes us so far because playing a DAW is improvising, composing, sound designing, arranging, orchestrating, recording, and engineering. This makes electronic music production a uniquely *omnimusical* experience, a mode of composing that engages sound from many angles and by many methods, a topic we return to in Chapter 6.

Principles of playing a DAW: feedback, interconnecting, working with waveforms, playing your musical system

The way producers play their DAWs as instruments is to *connect their music's sounds, parts, and effects to one another so they interact as a composite whole.* A helpful concept to illustrate this interaction is the notion of feedback. Feedback in the sense of one track plugged into another, interconnected, folding the music upon itself like its a mixing console deconstructed into a Cubist form. Feedback as returning this sound upon that sound, output becoming input becoming output again, in an ongoing loop. Producers play their DAWs by experimenting with feeding back their music upon itself at every stage of its making to hear what happens. It's through this process that

> the sample becomes the drone,
> the drone becomes the chord,
> the chord spins out a bass,
> the echo effect becomes a rhythm,
> and the rhythm modulates the sample.

Music production is a web of connected instruments, sounds, and parts. The producer plays the DAW by interconnecting as many components of the music as is feasible.

Earlier in this chapter we explored some widely used connecting techniques, including resampling, automation, and modulation. To these we add three more: side-chain compression, working directly with waveforms, and playing a musical system.

Side-chain compression. With this often-discussed technique, a compressor placed on one track is triggered into action by the audio output from another. Side-chaining, in other words, is a kind of machine listening. The classic side-chaining scenario is putting a compressor on a bass part that's side-chained to a kick drum. Each kick drum hit triggers compression on the bass which causes it to momentarily "duck" out of the way at the moment of each hit. This ducking not only clarifies a mix by making space for both the bass and kick sounds; it also creates a *dialogue* between the parts. Side-chain compression on the bass instructs it to "listen" to the kick, thus making the parts' interaction more musical—like a live band.

Working directly with waveforms. If you compose with MIDI, bouncing (or rendering) MIDI to audio or working with samples opens a fantastical rabbit hole of production possibilities. Producer Lewis Roberts (aka Koreless) prefers waveforms over virtual instruments because he has more control over their manipulation:

> Instead of using synths, I like to work with bare waveforms a lot of the time. I'll just record one long note and treat it in the Ableton arrangement view. I may move it and resample it and crop and/or reverse it. You can have better control over the envelope of notes and so on.
>
> *(Tingen, 2025)*

Recalling our discussion of automation earlier in this chapter, producer Joe Powers (aka Proc Fiscal) is fond of "stretching and manipulating" audio because "it feels a bit more tactile." Powers lets the shape of the waveforms guide him: "In Ableton you can just chuck it into the [sampler] and see the sound wave and how the sound wave looks. I kinda go off of that. Just seeing how it looks laid out" (Moss, 2019).

Producer Daniel Lopatin (aka Oneohtrix Point Never) finds opportunities in audio for discovering artifacts hidden within it. A producer, Lopatin says, should work

> destructively and think about twisting things up when they're audio. Don't always think synthesis is the be all and end all for sculpting stuff. I think audio is where it's at for me, because once it's audio I can really start twisting it into its final form.
>
> *(Future Music, 2022a)*

Producer Aaron Carmack (aka Mr. Carmack) advocates for working with waveforms as a way of moving from thinking about producing as playing instruments to producing as playing with sound's materiality:

> You can get past the instruments and the sound and get into actual technical audio wave files and shapes and forms. You can get really into the building blocks of what makes a sound a sound and if you can harness that power you can literally make anything you want...I think it goes beyond instrumentalism and beyond playing instruments and understanding core structure and musicality.
>
> *(Anagnostellis, 2016)*

Producer William Bevan (aka Burial) composes exclusively with audio samples. "I've seen people using sequencers and I've tried hard to use them" he says, "but it's blocks in different colors and I'm only used to just seeing the waves" (Fisher, 2012). For Bevan, the shape of the waveforms on the screen provides more visual information than MIDI blocks do. He's attuned to how audio *looks*. "I don't need to listen much to the drums because I know they look nice, like a fishbone, rigged up to be kind of skitty, sharp" he says (Fisher, 2012).

Since audio is captured sound rather than MIDI instructions for triggering it, working directly with audio is the closest one gets to *touching sound over time*. As producer Richard James (aka Aphex Twin) realized after having noticed the similarities between the waveforms of bird song and sine wave samples, audio files furnish visual evidence that "all sounds break down into really simple things" (Ward, 1993). Working directly with waveforms is a breaking-down-sound technique that allows the producer to shape it in endless ways, as if sound is Silly Putty. Interesting results come from simple techniques every producer can try:

> re-pitch audio into distant registers,
> granulize and stretch it into drones,
> find its affecting loops,
> chop it into percussive bits,
> use it as input for processing chains whose strange outputs will be input for even stranger audio.

Working directly with waveforms is a tactile way to maintain a sense of touch in production so that the music's elements are crafted, shaped, and adjusted just so.

Playing a musical system. The totality of your track's elements—its sounds, effects, MIDI instruments, audio samples, and structures—is a musical system that can be *played*. Learn to play your musical system as an instrument by spending time tinkering with it to understand the many things that can done with its elements. Follow sound engineer and visionary dub architect Osborne Ruddock (aka King Tubby) who played his mixing console "like an instrument with the faders and knobs like so many strings and keys he could use to transform the different parts of recorded material" (Fintoni, 2020: 24). Or follow producer Mark Summers (aka Interface), who noted in 2012, "I don't use synthesis, but I use [the DAW] Logic as my own synth" (Computer Music, 2012). *Your DAW is a synthesizer.* While this book can't provide a recipe for artistic practice, to play one's musical system as an instrument is to connect its many parts, unify its disparate energies, and blend its many outputs into a sound *you* find compelling.

Many lines from a single flowing gesture

Sometimes we fret over the direction a track is headed. We wonder, *Have I painted myself into a corner before I got going?* But we needn't worry—a direction can be figured out as we move along the production process. What's most important, initially, is our *gesture*. Gesture is the trick that makes thinking, feeling, style, and spirit manifest through our touch. Gesture is the artist's sketching, the cook's cooking. Chef Alain Passard equates gesture with the ways of the cook's dextrous hand: "Either we like the gesture, either we like the hand or we don't." It's through daily practice in the kitchen, Passard observes, "that makes the hand more precise, more accurate and more elegant. That's the trick" (Passard, 2016).

The composer Arvo Pärt memorably defines gesture as the nucleus or central core from which a piece of music grows. "I have always to find this nucleus first from which the work will eventually emerge…The compositional task is to find the appropriate system for the gesture" (Shenton, 2012: 117).[10] Pärt's concept is powerful. To build on it, we might think about gesture in spatial-physical terms—as the movements of our hands around a keyboard, for instance, or manipulating the controls of an instrument's interface. Urs Heckmann of U-he describes the gestures of using a MIDI keyboard as a performance by which "macros can be controlled from knobs and sliders on your MIDI keyboard. As the sounds can be played with dynamically changing colors, anyone can get creative" (Douglas, 2021).

Producer Grischa Lichtenberger performs his production process as a single gesture insofar that *he only works forward*, adding to what he has already done, but never deleting. This keeps him focused on the present moment and leads to interesting sound transformations:

> I very rarely go back or delete. I add one eq after the other and each one of them is like a step of correction or enhancement of the previous one. Of course it is maybe the worst thing you could do, but it makes the sound more yours, or better: it crystallizes what your relationship actually is. That's what I like about electronic music production—it isn't said that in the end the trumpet sound must resemble the original trumpet. After everything it could sound more like drums or like a piano.
>
> *(Headphone Commute, 2016)*

Think of a musical gesture then, as *a performance that unfolds over time*. What is *your* gesture? How does it move? How does it feel? Does it use variation or repetition? Does it keep your attention?

A composition often begins when we play or hear something that sounds like a part, an idea. But as producer Aaron Carmack pointed out above, gestures don't necessarily involve playing an instrument. A gesture can be a feeling, an intuition, an inspiration, a theoretical idea, or a techno-musical process. For example, maybe you've devised a way of organizing a collection of loops, or a way of applying an effect chain. No matter what its form, the defining attribute of a gesture is that it enacts a performance that compels us to want to build something with it, to organize music from it.

I now want to suggest the idea of *deriving many lines from a single flowing gesture*. This is a production approach by which a track's essential components are *generated during a single session in a continuous flow*. The approach is modeled on *ensō* ("circular form"), the Japanese ink painting practice of drawing a circle in a single expressive brushstroke (Figure 2.7). We conclude this chapter by considering seven production techniques for exploring and developing a single flowing gesture.

Improvise with an interesting sound and build on it. Play a part with a sound for a while, then tap your DAW's Retrospective Record button to hear what you did. Interfere as little as possible with what was recorded: if the part is 102.5 measures, work with it. Parts of unusual lengths lead to interesting forms. When producer Felix Manuel (aka Djrum) realized that playing piano *along to* sounds he had already programmed in the DAW wasn't working, he evolved a simpler strategy: begin by just playing piano, "just sit there and play and hit record" (XLR8R, 2018). Manuel found that playing for a long time produced "really natural

FIGURE 2.7 Ensō brushstroke: a single flowing gesture

performances" which he could sample at a later point. "I'd take big long improvisations, sometimes around an hour long, and then I'd pick out bits and make something out of that. The samples became the foundations for the tracks" (XLR8R, 2018).

By flowing forward like a brushstroke, big long improvisations remove some decision-making from a workflow, as if your gesture's declaring, *I have this today so let's use it.* "The improviser" notes Amir Chaudhuri in his book on improvisation in Indian music, "invents a problem in order to offer a solution" (Chaudhuri, 2021: 175).

Refine your first part before moving to a second. Usually a part needs refinement to bring it into focus. Even out volumes, fix obviously errant notes, or select the best moments from a wandering gesture.

Add a second part by deriving it from the first. Your improvisations contain *reams* of value to explore and develop. Derive a second line from the first part's MIDI or audio gesture. If the part is MIDI, borrow its notes to trigger another sound, as Ben Lukas Boysen explained earlier in this chapter. If the part is audio, convert it back into MIDI. This conversion process, notes producer Drew Daniel, "let you produce a kind of trace" of the MIDI. The unreliability of the process is an interesting telephone game, whose unreliability generates "a little grit of chaos" (Finlayson, 2022).

Ideally, the second part contrasts in some way with the first. This contrast can be in terms of timbre (e.g. a dry sound against a resonant first part), register (e.g. a bass against the first part's treble), or rate of movement (e.g. pulsating or changing at a different speed). Subsequent parts can be layered in a similar way, each sound attuned to contrasts. Contrast among multiple sounds helps them articulate and find a place in a track's arrangement and mix.

Once you have a part or multiple parts, resample them. Resampling parts is the surest way to turn a conventional sound into an unconventional one, creating new sonics that go beyond what you began with. Then make the music fractal by resampling a resampled part. Resampling has no end.

Reuse the effects or signal processing applied to the first part in altered form on subsequent parts. For example, if the first part uses an effect chain of two reverbs, reuse this chain somewhere else (and perhaps alter its settings). By pushing us to be inventive with one set of tools, reusing effects keeps our initial gesture flowing. Producer Jon Hopkins describes how he reuses effects intended for one sound on another:

> I'll do things like have a long chain of plugins I have saved in Ableton that have been built to process one particular sound, but then I'll try putting different sounds into that specific, really strange chain of plugins and see what you end up with.
>
> *(Mackintosh, 2021)*

Tinker with sounds until they sound compelling. As we have seen, sound in the DAW is always fluid. You can mute it, half-time it, stretch it, process it, re-sample it, re-pitch it, reverse it, cut it into smaller pieces, rearrange it, and so on, until the sound captivates.

In sum, cycling through these techniques bring us to a place very different from where we began with our improvisation. This journey reminds us that what matters is not *how* a musical idea originated or even how it sounded *initially*. What matters is the quality of the gesture, the quality of an idea composed or improvised or accidentally discovered and then set into motion. That's the trick: *trust the gesture*. Though its presence is fleeting, a single flowing gesture is enough to inspire the producer to make something from it.

Notes

1 A fine overview of the DAW and other music production software is Mike D'Errico's *Push* (2022). For a discussion of the Western music biases of this software, see Faber (2021).

2 For a discussion of Max as an aesthetic assemblage that problematizes the question of what a musical instrument is, see Snape and Born (2022).

3 Here is Booth's evocative description of Autechre's system: "We build systems because it's the most instinctive way of connecting things we invent. These systems do not produce a specific genre of music, they host modules, and manage all the protocols that allow them to communicate with each other. If we want to change the protocol, we have to enlarge the system. It was not more complicated than that. And the idea is absolute flexibility: you can change anything, anywhere. Inside the modules, it's more complicated: rather complex synthesizers, sequencers, effects with such parameters that they become a kind of synths too, with height controls, sequencers to control them…But this complexity is freedom. It's artistic. It's to create. I would be unable to make music differently today without feeling restricted. That does not mean that I can not appreciate music made from a cool synth and a great drum machine. But it would require such a mindfuck to come back to something so linear and limited. I like to compare our pieces to those Lego spacecraft we used to do when we were kids. You started from scratch and you stack the bricks with a very vague idea of the general design in mind. It just had to be almost symmetrical, and that it had wings. After, you improvised, you added elements, you removed others, and in the end you found yourself with a technological monster. We make monsters. The most beautiful possible" (Deshoqub, 2018).

4 As for working *inside* the computer, Frahm uses the DAW Cubase, but only as a recording device.

5 Buskin, 1995.

6 Raczynski continues: "It's always the same miswiring and corrupted brain config. The sound machines change but who cares about the sound machines, we don't need them, they need us" (Headphone Commute, 2024).

7 Andy Turner of Plaid again: "You might try and make a drum and bass track but don't have the skills so it turns out a bit weird...But that kind of weird is something new...something stranger and more exciting might come out it" (Gillham, 2019).
8 In 2010, a producer-fan on gearspace.com who uses the effects-laden sounds in Atmosphere, the predecessor to Spectrasonics' Omnisphere VST synthesizer, explained getting "tired of first having to switch off tens of FX sends and inserts to hear what's underneath" (Why I Hate Omnisphere, 2010).
9 Lynch, 2007: 1.
10 Pärt illustrates gesture's connection to the resulting music with the example of a struck bell: "If a single bell is struck, and we contemplate the nature of its sound—the Klang at impact, the spread of sound after this initial gesture, and then the lingering cloud of resonance—what we hear takes us to the heart of tintinnabuli" (Hillier, 1997: 20).

3

ATTENTION AND ATTUNEMENT

Introduction

Producer Sarah Davachi, who composes magnificent drone-based music using an assortment of vintage keyboards, offers us a perspective on production that gets to the principle of attention and attunement. Davachi explains that what makes a musician is the intensity of her listening—how she listens to sound in all its details, how she experiences their quality, feeling, and meaning. Davachi proposes a simple exercise: listen to the sound of a single oscillator. She urges us

> to really take time and listen...and not just move on to the next thing but really think about that sound quality. And you could try making minute changes to things and really try to absorb what that sound is like.
>
> *(Why We Bleep, 2020)*

Listening to the sound of a single oscillator is a way of distilling to its essence the listening acuity required of the music producer.[1] We listen to the oscillator to answer a fundamental production question, *What is this sound doing?*

The foundation of electronic music production is a quality of attention, a knowing how to focus on the endlessly variegated world of sound.[2] Underneath the producer's enthusiasm for gear (cutting-edge or vintage) and sound design is the under-celebrated skill of deeply engaging with what comes out of one's instruments, headphones, and monitors. This chapter explores electronic music production as phenomenological practice, illustrated with examples of how producers focus on sonic and perceptual minutiae in their work. It examines music production as phenomenology, hidden worlds in sample locations, the analytical-intuitive fader, play, musical process, getting each musical layer right, the untried, patience, and production-perceptual macros.

DOI: 10.4324/9781003585145-4

Music production as phenomenology

For the producer, being able to focus—knowing how to pay close attention to sounds' affects, suggestions, moods, and implications—is fundamental. Paying close attention is a production superpower because it positions us to perceive the details that make the difference between a conventional sound that conveys little and an interestingly alive sound that conveys a lot. Attention is the metaphorical wavelength upon which producing operates: find the right frequency, get the clear reception. Davachi's suggestion to *really think about that sound quality* reminds us that producing is attention and time intensified, a practice of immersing oneself in the space of sound that conjures counterpoints of feeling. Producer Brian Eno, who we heard from in Chapters 1 and 2, describes his ambient music practice not in terms of a preference for gentle, long-sustaining sounds, but in terms of putting himself into a space of attention. "I'm addicted to the idea that you put yourself in a place and surrender to it" he says. "It's about making space for a kind of attention that you're not normally offered by entertainment media" (Marchese, 2022). Eno is not alone in his view of production as a means of putting oneself into an attentive state. Let's consider a few other examples. Producer Mark Fell frames creative work as attunement:

> I want to promote a description of creativity as a process of attunement to the material environment, not an isolated or inward journey further into one's thoughts or mind or soul. In this sense, the description I want to promote is one driven by a critical curiosity rather than a thing called inspiration.
>
> *(Fell 2022: 21)*

For producer Cephas Azariah, there's a connection between foregrounding the details in ambient music's small sounds and the careful, attentive mindset these sounds engender in a listener:

> For me, ambient music is drawing out the details in the small sounds. Whether that's a piano, a synth, orchestral textures or a pattern, it's about bringing those elements to the forefront of the sonic space. Making this kind of music requires an intuitive slowness and attention to detail. For example, you could gain up the mics on a felted upright quite a bit to hear the mechanics, the actual notes being played, the room, the piano stool, the player and really draw the listener in as if they are sat next to you at the piano. It's just one instrument, but that's sometimes enough in ambient music.
>
> *(Payne, 2023)*

Producer Sibel Koçer (aka JakoJako) understands her production process in the studio as playful. But equally it's a meditation tool for making hypnotic music conducive to changing her perception:

> Making this hypnotic, shifting music comes naturally to me when I'm in the trance of the studio. When I make music, I enjoy it myself a lot, it's for me a kind of meditation tool. No pressure, just playing with synths. Maybe that's why it sounds hypnotic, because I was hypnotized too.
>
> *(AIR, 2024)*

Another example is producer Moritz von Oswald, one half of the duo Basic Channel (with Mark Ernestus), whose dub-influenced music engages attention through a minimum of sounds, subtly interacting. In the 1990s, Basic Channel's minimal techno was built from just three or four elements: a muffled 4/4 kick drum, a sub bass, and a single guitar or keyboard chord rhythmically fluttering through delay effects, echoing above the pulses. Like the best dance music, Basic Channel's is repetitive, yet kinetically alive and forward moving. Von Oswald characterizes this sound with a concept he calls *steady state:* "Steady state. Something which is not really standing, but standing and moving at the same time. Then it leads to a really deep state" (RA Exchange, 2023). He explains the duo's techniques for achieving this deep state:

> We used patterns that weren't really defined, but they work with each other and are constantly changing. It's a state where the music is not moving, just kept like a Polaroid, in a way, where you have something going on rhythmically, but it actually stands still at the same time.
>
> *(Schmidt, 2008)*

Like Davachi, von Oswald urges producers to use their production process as an opportunity to pay attention—particularly, in von Oswald's case, to what's in the background of one's tracks. Producing, he says, is "experimentation, exploring, sound, and listening. To listen to backgrounds. You should always be listening to what is in the background, of whatever it is...I'm trying to round up the capacity of what the ear can perceive" (RA Exchange, 2023).

Building on Davachi, Eno, Fell, Azariah, Koçer, and von Oswald, we might think of producing as a space for pushing our capacities for noticing within musical experience. *Really think about that sound quality, make space for a kind of attention, get attuned, draw out the details, get hypnotized, round up what the ear can perceive.*

Consider that to produce music is to practice phenomenology, a philosophical approach oriented around the study of consciousness and the objects of our direct, lived experience. It's the quality—the affective feel—of this experience that's the subject matter of the phenomenologist.[3] Philosopher Arnold Berleant vividly describes what the space of music feels like while we're making it: "The performance situation catapults a musician into a rare and unusual condition" he says, "one that reveals the basic features of experience with eloquent directness, free, at least to some extent, from the usual overlay of cultural and philosophical presuppositions that nearly always obstruct our awareness" (Berleant, 1999: 75).

When the producer puts on headphones or sits in front of the studio monitors to work, ideally there are no presuppositions, only the listening condition in which a world of feeling opens up, catapulting her into a stereo space the experience of which feels uniquely immersive, powerful.

To help further frame the basic features of the producer-as-phenomenologist's experience, there are two examples of phenomenologically-oriented writing that inform this book. The first is *Ways of the Hand* (2001), an investigation of learning to improvise jazz piano, by the late musician and sociologist David Sudnow. The text describes the learnings of a pianist's hands, mapping the paths they take along the terrain of the piano to distill "the concrete particular problems" the jazz improviser faces and to show the thinking involved in musicianship (Sudnow, 2001: 3). Sudnow's reflective writing brings us into how he perceives the keyboard world. "How the paths sounded to me was deeply linked to how I was making them" he writes. "There wasn't one me listening, and another one playing along paths" (Sudnow, 2003: 40).

A second example of phenomenological writing is *Becoming Animal* (2010), a book by ecologist and geophilosopher David Abram, which describes an education in becoming receptive to the moods, rhythms, and affects of nature's non-human inhabitants. *Becoming Animal* is a guide to perception that documents the author's fluid relationship with nature and its writing resonates beyond its subject matter to catapult us, like Berleant's musician, into the phenomenology of experiences. From the shadows and depths of mountains, the moods of weather, to the sleights of hand of magicians, *Becoming Animal* conjures the energy and dynamism of Abram's encounters in prose that's *musical.* Mountains, for example,

> beckon to us from behind the cloud of words, speaking instead with gestures and subtle rhythms, calling out to our animal bodies, tempting out skin with their varied textures and coaxing our muscles with their grace, inviting our thoughts to remember and rejoin the wider community of intelligence.
>
> *(Abram, 2010: 40)*

Similar to how mountains speak to Abram through the subtle rhythms of their presence, so too do music's soundscapes beckon the producer. Producing is phenomenology because it brings its practitioners, on a granular level, into music's terrains of textures, dimensionalities, and forms.[4] Whether we're playing an instrument or mouse-clicking around a DAW, our producing triggers a multiplicity of sensations. Our minds move far and wide—feeling and associating, conjuring and remembering, analyzing and imagining, connecting our beats, chords, melodies, and timbres to sensations, times, people, and places. In sum, the pioneering Deep Listening composer Pauline Oliveros captures this multiplicity of sensations, this omnimusical experience, when she says that "to listen is to pay attention to what is perceived, both acoustically and psychologically" (Oliveros, 2005: 22).

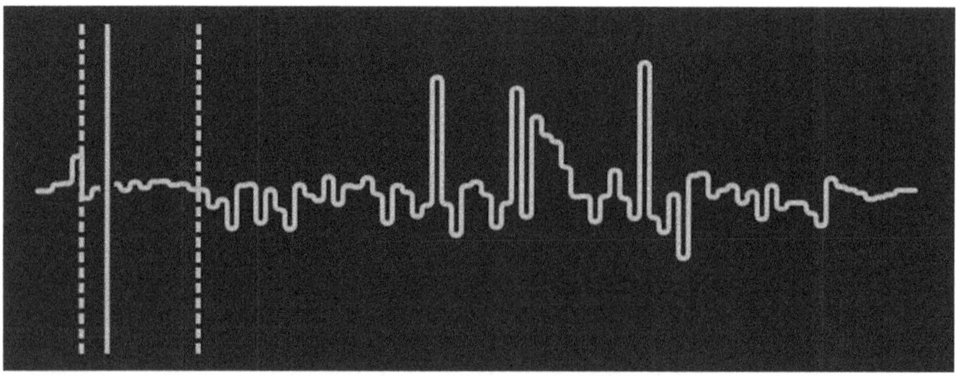

FIGURE 3.1 Audio waveform in Fors' Opal sampler-sequencer

Hidden worlds in sample locations

Sometimes in samples of one's earlier work there are hidden worlds that could be points of departure for a new track. I load an audio file into a sampler-sequencer, posing the question, *Is there something here to work with?* The sequencer's small sample window displays the jagged peaks of the file's waveform, compressing three minutes of sound into two inches of what resembles an abstracted city skyline (Figure 3.1). I click on the first step of the 16-step sequence and listen to the bit it triggers from the sample. The sliver of sound is barely long enough to be audible, so I extend its sustain and release to make more of the sample play each time around the sequence. But the instrument's most important control is "Sample Location", which allows for scrolling through the audio file to select different loop start points. Each sample location is a casting of a hook into sonic depths. I nudge the cursor around and listen, wondering where the best location is. There may be many options that sound good, that could work, and I want to try them all! But the best sound is often *what's most captivating right now*. Our captivation is as important as the sound itself. After all, "the things we care about are the things we make art about" write Brian Eno and Bette Adriaanse. "We frame them with our attention" (Eno and Adriaanse, 2024: 83).

My search for the best sample location illustrates production as a phenomenology of one's own experience through sound in the search for sensations we can't encounter otherwise. The producer's interactions with sounds is a quest for what the anthropologist Robert Plant Armstrong calls an *affecting presence*, an aliveness at many levels at once with a renewable power to summon feeling (Armstrong, 1971). In *The Affecting Presence* Armstrong drew on examples from Yoruba and Javanese expressive culture (sculpture, music, dance) to propose that we understand artworks not merely as symbols or representations of social life, but rather as direct "presentations" of feeling and affect.[5] Armstrong's phenomenology of art as an inherently affecting presence aligns with how music producers experience making sound. At every moment of making music, the producer-as-phenomenologist assesses how the sounds in a track-in-progress feel. Feeling in music, a topic of Chapter 5, is a composite thing, the result of many interactions. Consider, for example the Amen break, the legendary drum solo sample from the Winstons' 1969 song, "Amen, Brother". For producer Karl Nicholas Francis (aka Dillinja) what makes the break feel perfect is how all the components of its sound combine. "That's why the Amen is a perfect break. The textures, the grains, the air of the [mixing] desk, everything about those drums is magical. It's got its own soul" he says (Jenkins, 2020).

Whether there's three sounds or twenty, and whether those sounds are acoustic, synthetic, or in between, each has a unique presence. A sound can be mellow or strident, clinical or plaintive, distant or near, razor clear or obfuscated, and the repeated production task—so obvious as to be unremarked upon—is to find the right sound for the musical context and the right musical context for the sound. So I continue scrolling through the sample locations. I'm searching for a sound that sounds alive; listening for an affecting presence that sounds, like the Amen break, *magical*. But what I hear feels too energized. I pitch down the audio by six semitones. That's better; lower-pitched audio is more mysterious. A promising sign is that the sample's original source is fading—I no longer recognize where it came from.

Maybe in this lower register I'll hear something I can work with? Putting samples of one's earlier works into a sampler/sequencer is an always revealing resampling workflow because we don't know what we'll find. As the sequencer cycles around, I listen for a point at which the sliver of audio benefits from repetition. I play with the settings at hand—changing the attack and decay so the sample fades in before slowly dissipating, filtering out some of the highs, and adding reverb so it lingers in its pitched-down resonance. Sometimes the sum of a few small tweaks is enough to get a sound sounding like it was meant to be.

The analytical-intuitive fader

There's a volume mix fader on one of my virtual instruments that allows crossfading between the outputs of two oscillators (Osc). The control is simple: when the fader is all the way up the sound is 100 percent Osc 1, when all the way down the sound is 100 percent Osc 2, and when positioned in the middle it's 50 percent of each (Figure 3.2). This is the kind of easy-to-use parameter I turn to when I want to shift a sound slightly, but without making it unrecognizable. Subtle sound design—whereby one minutely transforms a sound in a way that's felt more than noticed—is often the most effective kind. Producer A.G. Cook takes this approach to using one of the dials on Lazerbass, an early monophonic bass synthesizer by Native Instruments. Cook explains how he uses the instrument's main dial to morph a sound into variations of itself:

> Because that's on a dial, you've got something harmonic on one side and something sort of percussive on another side...You've got a dial that goes between those two points. So there's already a whole composition exercise or sound design tool suggested.
>
> *(Tape Notes, 2024a)*

Zooming out from sound design to mindset, the Osc volume mix fader on my synthesizer is a metaphor for how producers shift back and forth between different modalities of

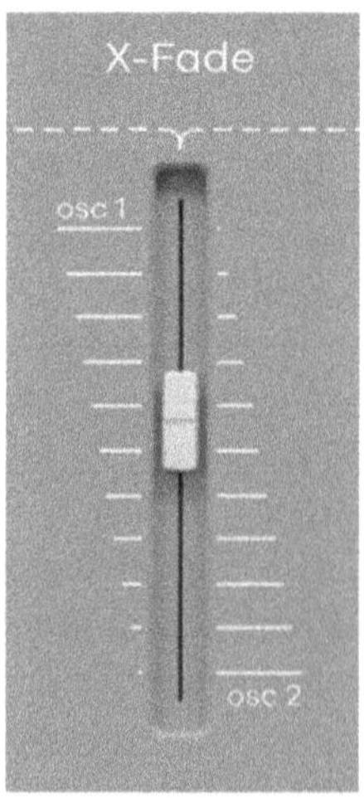

FIGURE 3.2 Oscillator 1/Oscillator 2 mix fader

knowing and doing. These modalities are analytical and intuitive: the analytical way is a conscious cognitive activity, while the intuitive way is unconscious. Both modalities are in play as we adopt different frames of mind whilst making decisions about what and how to create. In my work I begin by consciously and deliberately committing to a sound, usually a quite simple one. Producer Seth Horvitz (aka Rrose) describes the "narrow field" approach to production, which prioritizes sound over personal expression, especially at the outset of a project. "The approach changed from putting personal expression first to discovering things about sound" Horvitz says. "Now I consciously work with very basic building blocks of sound" (Smith, 2020).

Next I try to do something with the sound by finding a kind of performance. A performance can be a minute or an hour, but is has to achieve something, it has to *go* somewhere. Where the music goes doesn't matter as much as the performance *sounding compelling where it is*. A producer's performance is arguably a track's most potent ingredient—the gesture that contains energy, as we discussed in Chapter 2. I play the sound using the keyboard, intuitively reaching for notes as shapes and patterns, lingering on chords that are interesting, trusting the hands to lead me along paths through remembered symmetries.[6]

But once I have a performance recorded, I'm back in analytical mode for a moment, trying to figure out what happens next. Do I need additional sounds? If so, will they be variants of the first sound, or not? At any stage of a workflow there are so many ways one can go. I try out sounds and parts—copying, pasting, moving, stretching, and resampling to scale up and fill out my performance. Producer and pianist Neil Cowley develops and embellishes the ideas in his piano playing through a gradual, additive process he calls orchestration:

> Piano first, generally. Though maybe a small rhythm element may be put in place to spark the harmonic gateway that the piano provides. Then starts the long process of embellishing it, or orchestrating it. I say orchestrating, as I like to think of the synth elements that I add as orchestrations.
>
> *(Headphone Commute, 2021)*

This trying out of sounds and parts is itself another kind of performance, an improvising on methods—I try whatever occurs to me, going with what works then moving onto trying another bit. The process zig-zags, from one sound layer to another, from analytical mode to intuitive mode and back again, as I embellish the sonic whole in a way that enhances it. In sum, in the space of a single session a producer may shift numerous times between different modalities of knowing and doing. Like dragging the synthesizer's mix fader between Osc 1 and Osc 2 to re-shape a sound, alternating analytical and intuitive mindsets is a way of responding to music's subtle rhythms and varied textures in the moment.

Play

To produce music is to play with sonics by devising sounds that have affect, that do emotional work, that hold our attention. But what holds our attention is not just specific sounds per se, it's also what the producer *does with* sounds. Skilled producers reveal sounds' potentials by playing with their artifacts, taking them apart, and rearranging them into new constellations. The best producers are kaleidoscope-makers who do novel things with often quite limited soundsets.

How does one play with sounds? A first step is to start simple. Use sounds that are either generic or roughly in the area of what you want and then process to alter their timbre or their part until these become unfamiliar, *I-didn't-know-I-would-like-these* sounds. Producer Dominic Angus (aka Dom & Roland) suggests getting one's sounds

> in the ballpark and then start spanning it out as quickly as possible...Then you start picking away at various bits and say, 'I'm going to process this bit because this feels exciting at the moment.' I suppose the whole thing is about keeping the excitement alive for you as an artist while you're being creative.
>
> *(Computer Music, 2017)*

A producer's play is mangling these sounds into something other, something not heard before that keeps excitement alive. One can, of course, go overboard with sound mangling. Drowning a sound in reverb, for example (a classic misunderstanding of ambient), tends to obfuscate the sound's source. On the other hand, some effects like distortion work well on max. Distortion can pixellate a sound into a thousand noise pieces. Producer Matthew Barnes (aka Forest Swords) uses it to "create really gnarly sounds that have been mangled so hard you can't really tell what the original sound was and can pretty much create new instruments or something you've never heard before" (Turner, 2023a). Low- and high-pass filters also work well at extreme settings: a heavily low-passed sound is ethereally mellow, while a heavily high-passed sound is brittle like glass.

After processing sounds, a third step is rearranging them. Sometimes a sound's part has an evident, obvious structure—for example, an eight-chord sequence, or a four-bar repeating kick-snare pattern. These repeating sequences and beats suggest obvious arrangement shapes. But chords or beats also contain less obvious latent structures: the producer can play with a triad that contains yet-to-be-explored harmonies ready to bloom, or a looping breakbeat that contains other rhythms hidden inside smaller segments of it. Producer Jan Jelinek, whose work with loops are explored in Chapter 8, sees his audio editing as a way of revealing such insights within sound. Slowing audio down is a favorite technique. "The most simple ways of audio editing" he says, "are naturally the most effective ones: deceleration and acceleration. Especially deceleration is a wonderful way of microscoping analysis. It exposes rich and abstract textures full of former hidden information" (Fosl, 2018). We discover new harmonies and rhythms when we playfully unpack the possibility spaces of our chords and beats.

In sum, production play is exploration. The chord sequence or the beat always has more to offer, and our production adventure is to *uncover the latent structures that exist beneath the obvious ones*. Re-order the chords and edit their notes; slice the beats to reveal syncopations within the syncopations. There's no one best way to approach rearrangements, but it's always playful. Try things, listen to what sounds good, and build on it. Play with what you already have to try to make something you haven't yet heard.

The main driver: musical process

The music streaming platform Spotify offers artists an opportunity to pitch their music before its release for inclusion on one of the service's many curated "editorial" playlists. The pitch involves providing information about the track's instrumentation, mood, culture, and which three styles best describe its sound. But the most important part of the pitch is explaining your *process* of composing the track and what *inspired* it. This narrative is allegedly considered by Spotify's editors as they decide whether or not the music deserves to be on a playlist. The point of this story is that a common way to think about music production is in terms of a musical process/musical inspiration dichotomy. Process is what producers actually *do* in terms of their workflow, while inspiration is what they *believe* guides their doing. Such framing of artistic practice in terms of inspiration communicates what we *want* the work to be about, not necessarily what it *is*. But inspiration and process are linked experiences. In fact, fruitful musical-technical processes often become sources of inspiration.

Surveying interviews with numerous producers, we notice than many of them work out of curiosity rather than inspiration, in the sense that it's by making musical stuff that they answer the question of what to make. In other words, *one's process of making becomes one's musical inspiration*. To return to the Spotify pitch, process talk is perhaps not as poetic as inspiration narrative, but it is a more accurate accounting of how producers compose. Process is music production's through-line: one's workflows within a musical system are what generate materials for new music. In fact, when producers talk about making music most of their ideas concern details of process rather than inspiration's vague sources. With this in mind, let's now consider how musical process and musical processing are workflow's main drivers.

Producer Chris Clark (aka Clark) describes his work as "a process of incremental discovery everyday" (Spitfire Audio, 2018). While its daily results wax and wane and require curating, Clark trusts process to generate ideas that will ultimately coalesce towards a musical goal:

> Not every day is amazing, but being able to work through that and just accept that one day you're going to discard things but they're all going towards something. It's like a taste thing—it's like a sensor in your worst impulses. So perhaps [process] is underrated.
>
> *(Spitfire Audio, 2018)*

Producer Greg Jones (G Jones) describes his workflow as an *exploratory* process:

> The more you fuck around and try new things that seem like they won't work, the more likely you are to discover some magical combination of sounds or effects or whatever that you would not have discovered if you were trying to achieve a specific sound as opposed to having an exploratory approach.
>
> *(Kraker, 2018)*

Producer Jimmy LaValle (aka The Album Leaf) understands process as a way of sound design experimentation that generates inspiring results:

> Sound truly gets me inspired. It's what I react to. A patch I created, then sent through this FX chain, then sampled and put into [Ableton] Simpler and back out through another FX chain. That method of experimenting inspires me constantly.
>
> *(Headphone Commute, 2023c).*

For producer Chihei Hatakeyama, sound processing is the first step of a two-step compositional process:

> My compositional process can be broken down into two steps. The first is processing the source material. The second is determining what to do with the prepared material. Using computers and outboard gear, it is possible to process material in an infinite number of ways, so you could say that processing could go on forever.
>
> *(Fischer, 2012a)*

Producer James Holden, who we heard from in Chapter 2's discussion of musical systems, understands process as an accidental generator of songs. Holden relies on process to bypass his preconceived ideas of what the music should sound like:

> I usually sneak up on myself, like I start designing an instrument and then the songs come out by accident as a result. It's weird—sometimes I'll go in the studio and something will come out by itself, but the process and the method of it is really important to me.
>
> *(Mullen, 2023)*

Producer Richard Devine uses process to discover unusual sounds. One of his techniques is to use impulse response files of acoustic instruments (that capture their acoustic characteristics) to modulate other sounds through convolution (the process of multiplying the frequency spectra of two audio sources):

> I will use objects, like cymbals, mallets, rhythmic pulses, or granular textures and then use the impulse file to impose the amplitude and frequency of that sound and convolve it to the incoming audio. I have discovered so many interesting sounds with this process.
>
> *(Headphone Commute, 2014a).*

Like Hatakeyama, producer Geir Jenssen (aka Biosphere) uses sampling and processing as the core of his compositional process. To make his 2012 recording, *L'incoronazione di Poppea*, Jenssen manipulated small samples—what he calls *fragments*—of Monteverdi recordings:

> Even in the approach to Monteverdi, I worked on the fragment, I would say even on the fragmented nature of the creative process: I worked on sound tissue sampling and processing an enormous amount of material taken from all releases, on vinyl and tape.
>
> *(Napolitano, 2010)*

Producer Felix Manuel (aka Djrum) characterizes the workflow to make his 2024 recording *Meaning's Edge* as "you're problem-solving as you go, but it's more exploratory...I'm trying stuff out and kind of seeing what works" (MusicRadar Tech, 2025). Finally, producer Ryan Lee West (aka Rival Consoles) trusts his daily process of exploring sounds, whether or not he ever uses them in a track:

> I'm constantly exploring stuff on different scales and then the stuff that interests me ends up on different compositions, potentially. But even if things don't get used, often you learn from them, or you learn what you like or don't like, [from] the process.
>
> *(Why We Bleep, 2023)*

In sum, producers' varied processes remind us that inspiration isn't always the main driver of new music. Questions such as Spotify's *What inspired your track?* miss the fact that sometimes the most inspirational source of ideas is the process of making itself. Over and over again, producers use workflows to discover music by creating conditions for sounds to flourish and then tinkering away within this environment. *Process, process, process.*

Get each musical layer right

To explore process more, let's consider a production strategy that assures musical results: getting each layer of a track-in-progress right before moving on to the next. First, we look at what a musical layer is, and second, what getting it right entails.

Each component of a track constitutes a layer of the music's total texture. Typically (and to generalize), an instrumental electronic music track has at least six sonically distinct layers which may include variants of beat/percussion, bass, chords/pads, lead/melody, Foley/field recordings, and effects. Some tracks have many more or fewer layers, but variants of these sound categories are often present.[7]

Beat/percussion. Freed from the drummer's labor, electronic beats can take many forms and can be constructed out of any sound or collection of sounds. But beats' most common sources are recognizable drum and percussion instruments, as well as drum machine approximations of drum sounds—for example, the Roland TR-808's famously un-cowbell-like cowbell. A beat will often include a low frequency kick drum-type sound, a mid-frequency snare drum-type sound, and high frequency sounds, like hi-hats. These percussion sounds usually have sharp attacks and short durations. But producers also make drum sounds out of white noise, field recordings, voice samples, or anything really. Producer Josh Davis (aka DJ Shadow), who made his 1996 debut, *Entroducing*, using just an Akai MPC sampler, explains that having numerous options for making a kick drum has shaped his *sound is just sound* philosophy of production:

> "If I'm making a song and I need a kick drum, well, I could sample a kick drum, I could use a vintage drum machine emulator, I could mic up a drum kit and record a kick drum...At a certain point sound is just sound. It really just comes down to what

qualities you are looking for from the sound. That's really the way I think about music at this point."

(*Mullen, 2023b*)

A beat, however constructed out of whatever materials are at hand, is often a track's anchor because it articulates tempo, rhythmic pattern, and feelingful groove all at once. This is why (speaking as a percussionist) drumming, even when freed from the drummer's hands, is music production's most interesting constellation of energies.

Bass. Along with the kick, a bass part is the main provider of a music's low frequencies. Synthesized bass tones and bass guitar samples work, and so does a sine tone pitched down into the bass register. Bass bring depth and richness to music. Nothing is more substantial than a sub humming several octaves below the music's surface, doubling a mid-range chord's root note. Bass's slow-moving frequencies envelop us like ocean currents.

Chords/Pads. Chords are a track's glue that connect bass and melody parts. They also connect us to the long history of tonality from European classical music that continues to shape electronic musical practice. Our chords tell a subliminal story over time through their progressions, suggesting to the listener that this feeling is jubilant or foreboding, or any of a myriad *it-feels-sort-of-like-this* sensations. Producers typically play chords on keyboard or grid controllers, or else program them, a practice that can lead to unusual harmonic structures. Producer A.G. Cook works this way, composing chords by drawing in MIDI notes in his DAW's piano roll editor. Discussing how made his track "Heartache" (2024), Cook explains the options afforded by drawing MIDI:

The sort of harmonic style I have is very much based on being able to flip the octave, mess with the intervals, inversions, all that stuff, and see things quite visually as well as listen to them. So, that was a sort of MIDI drawn-in loop I would have [drawn].

(*Tape Notes, 2024a*)

Lead. A lead is most often a track's melody which, in instrumental music, evokes the expressive singing voice. Not all tracks have a lead, but when present the lead is the music's focal point.

Foley/field recordings. As we'll explore more in Chapter 7, producers use Foley and field recordings, processed or not, to bring the depth and realism of an acoustic world to a track. The workflow of producer Joseph Kamaru (aka KMRU) involves playing parts over his field recordings. "I will drop a long piece of recording and improvise and layer on top" he says (Housego, 2020). Field recordings are also incorporated into effect plugins such as Devious Machines' *Texture* and SoundGhost's *Texturize*. These effects offer snippets of field recordings as textural sources to layer onto audio or MIDI, enabling a producer to, for instance, add running water sounds to thicken the transients of snare drum hits. Field recordings are also a source of tonal material. A producer might use an audio-to-MIDI tool to transform a wind chime recording into MIDI, then use this MIDI to trigger a chord part.

Effects. Effects include conventional devices such as reverbs, compressors, EQs, as well as bespoke effect racks that process sounds in wild ways. Producers often begin with an unadorned conventional sound and then use effects to transform it into something filigreed and unusual. The powers of sound transformation give the producer confidence that workflow is less about where we *begin* with a sound than where we take it *to*. As producer Guy Brewer (aka Carrier) explains, processing a Roland TR-808 tom sound can take it far afield. "I really like the way drums sound when they are pitched beyond their normal realm" he says. "I'll have an 808 tom, but you won't really be able to tell that's what it started life as" (XLR8R, 2017).

These brief descriptions of the nature of different layers in an electronic music track, while generalized, offer an overview of the sound types producers work with. Two additional things to remember: a complex-sounding track may only have a few layers, and conversely, a simple-sounding track may have many layers.

Now we turn to the idea of getting each layer right before moving on to the next. This has two components. First, create a sound and a part for that sound that are enchanting on their own. Second, commit to both the sound and the part.

Let's illustrate this by way of making a beat and some chords to go with it. You might begin with a tempo, choose an initial sound set, and build a simple kick and snare pattern. (I often just drum something using an 808 kit to establish a rhythm and tempo, without worrying about timbre specifics.) Over this kick and snare you could layer hi-hats or other high-pitched percussion parts. You might re-pitch and quantize elements of what you've played, then double the pattern's length.

Next you can experiment with applying effects like delays or compressors to individual drum sounds within the beat, so that each sound hits in a unique way. (Note: This may *radically* change the rhythm you began with!) Then maybe you'll mute drum hits here and there to make space to better hear the effects. Continue tweaking the beat's parts and effects until their composite sound coheres and their rhythm moves— aloft on the energies of its groove, self-propelling around and around. "Getting an interesting, evolving rhythm section is always the goal" notes Brewer. "In my opinion, the way you handle and mix your drums is one of, if not the most important skill to master" (XLR8R, 2017).

You might try pushing the beat further afield, processing it with a multi-parameter effect that adds layers of motion, glitchy stuttering, and the subtle nonlinearities of magnetic tape saturation. If this produces enchanting results, process again, running the beat's sounds through another layer of uncertainty. The goal of such *multi-layer processing* is to explore ways of transforming what began as a simple beat into a one-off sonic object. Push yourself to push the sounds. Producer Nolan Petruska (aka Frequent) explains the relentlessness of processing. "A lot of times people ask me how I made a sound, and it's like, *I just didn't stop where you would have stopped twenty effects ago*" (Live Producers Online, 2023). By trying out effects and finessing their parameter settings (especially the wet/dry knob which allows each effect to be used subtly, like a tint) you may discover a fascinating rhythm with just the right amount of strange.

With a strange beat in hand, move on with chords. Chords can be used in a number of ways. One way is to sketch a tonal journey for a track: a chord sequence (repeating or not) gives the beat layer directionality. Chords also establish a key for the piece; even a single chord or sustained note can be a placeholder drone upon which to build other parts. And chords can be used to push against the sense of time set up by the beat. An interesting perceptual effect is slow-changing chords over a fast beat, a contrast that generates a meditative feel.[8] As with your beat, take time to craft a chord sequence that compels you. It could be brief but energized, or long and never quite repeating. Ideally the chord sequence sounds good on its own.

Getting each layer right before moving onto the next isn't the only way to work. Sometimes producers quickly sketch out each layer, then slowly finesse their parts later. Here is A.G. Cook again, suggesting exactly this:

> My preference is really to do a skeleton of a track with quite bad sounds—or not bad, but underdeveloped sounds—and know that the bones of the song are decent. And then I'm confident I can go in and mess around to make it sound cool.
>
> *(Tape Notes, 2024a)*

The power of the *getting each layer right before moving on to the next* protocol is that it generates parts that work well together and individually. This makes arranging a track more manageable. For example, if you decide that the entire beat layer of your track will be muted in the final arrangement, the sounds that are left will still have impact. In sum, getting each layer right focuses the producer's attention on making a sound interesting in itself before other layers join the mix.

The untried

Sometimes the same workflow habits that produce results keep us in ruts, because returning to proven techniques forecloses discovering new ways of doing musical things. The more techniques we explore, the more we realize how much we've yet to explore. How then, do producers draw on what they know and continue to innovate? Consider three approaches to explore the untried.

Seek new timbres by trying new combinations of sounds and effects. We're familiar with how combinations of conventional acoustic musical instruments sound because they've been used for a long time; these instruments have histories of playing together as ensembles that go back hundreds of years.[9] But we know less about how blends of electronic timbres sound, in part because electronic music's history isn't as long. Each track then, is an opportunity for a producer to try new timbre combinations. Consider a pristine bell with a fuzzy bass, or an icy ambiance over a warm pad. Build a world entirely from field recordings. Or combine effects in novel ways: mix wide stereo with narrow mono, connect reverbs to make a gargantuan hybrid space, blend the pristine bell with the fuzzy bass to make a fuzzy bell-bass. Trying new mixes of sounds and effects can be a daily practice. Each novel timbre combination is an opportunity to create a new meta-instrument, opening a sound design journey without end.

Amplify and scale up what you already have. The moment that jumpstarted a track may have been brief, but you can extend this moment by scaling it up. Develop the chord sequence not by repeating it but by playing around with it—altering the pitches of its notes, or stretching/compressing its length. If a part's sound isn't enchanting yet, use the part to trigger another sound that is, or resample it to generate inspiring audio.

Be intentionally naive. At any point in a workflow pursue the most compelling thing that catches your attention *before* you understand how the thing works. When you come across an interesting sound, move quickly to build something with it; make variations *immediately* to hear what it can be, to get a sense of its potentials. As Matt Ridley observes about innovation across scientific and artistic fields, "techniques and processes are developed that work, but understanding of them comes later" (Ridley, 2021: 282). Musical building—not theorizing—is both a producer's way forward and how production itself evolves.

Patience: nothing is happening, until something is

Composer-producer Nils Frahm, who makes music that incorporates meditative keyboard improvisations, dub aesthetics, and extended structures, allows himself time to get into a performance when he's recording. "To give the moment a chance, you don't give up after three, four or five minutes of playing" he says. The goal is not to "play the peak moment in the first three minutes" but rather to give the music time to bring oneself "deeper into the zone" (Mullen, 2022a).

Producing music requires patience to let conditions coalesce, time to give musical moments a chance to unfold. Sometimes this is difficult to do because it feels, for the longest time, as if nothing's happening. We might have begun a session with an expectation that we should have an interesting sound or groove happening within a few minutes. But instead we fumbled around, wondering what to do and where to go, feeling exasperated. We play some chords with a sound, but are unimpressed with both the chords and the sound. We search for a way through the impasse as time passes: we're trying different things, but they're not working. Are we, as Frahm says, *giving the moment a chance*? Sometimes it helps to simply wait for something to happen and be receptive to opportunities that may present themselves. In his book *Wild Problems*, Russ Roberts advises this. "Sometimes doing your utmost means merely waiting. But it's not waiting idly. It's waiting attentively" he says. "It's about paying attention. Sometimes doing your utmost means doing nothing but being ready for what comes next" (Roberts, 2022: 184).

Waiting for something musical to happen is a transitional mindset between being in a state of expectation and being in a state of receptivity. Searching for ways through production impasses keeps us in an attentive zone, alert for anything interesting that may appear. Sometimes searching—

Can I pitch this down?
Can I filter it?
Can I swap one sample for another?
Maybe this sound is enough?

—opens the door for a musical answer to present itself.

Producers spend hours trying techniques for making interesting sounds, an investment of time and energy that illustrates how the craft asks for a commitment to process without guaranteeing any returns. The longer nothing seems to be happening the lower our expectation that anything significant will *ever* happen. Waiting can feel bleak, but herein lies a gift: lowered expectations open us to a heightened state of receptivity. As Frahm elaborates on his approach, "you're just getting to a point where you're not disturbed by your own bad ideas, being very humble" (Mullen, 2022a). Ironically then, tinkering around in a production workflow with diminished expectations that anything significant will ever happen creates the conditions that spark ideas. Why? Because without expecting to notice anything specific, the producer is better positioned to notice. In sum, give yourself and the music a chance. Listen and pay attention to what you're hearing. For what feels like the longest time nothing is happening, until something is.

Production-perceptual macros

The term *macros* comes from computing, where it refers to a single instruction that unfolds into an instruction set for carrying out a task. In music production, macros are single commands triggered by a control (e.g. a knob) to carry out several tasks at once. In the DAW Ableton Live, the software's clever macro knobs on its effect racks allow a musician to control multiple parameters on multiple devices within a rack *simultaneously*. For example, one can map a macro knob to the dry/wet controls on a reverb, a filter, and a distortion. Turning the mapped knob increases or decreases settings on all three effects at once, creating a composite effect. Recalling what Guys Sigsworth said about automation in Chapter 2, macros enable us to have eight pairs of virtual hands working on multiple parameters at once. Macros extend our senses and amplify our touch into multiple dimensions of sound-feeling.

We can also think about macros in a less technical, more generalized sense—as suggestions producers offer themselves on the way to furthering and finishing tracks. Journalistic self-help articles about music production teach workflow techniques and hacks for getting musical things done. But such well-intentioned advice has limited utility because each one of our tracks is its own uncharted terrain, a topography-in-progress without a readymade map, a unique aesthetic situation that requires all of our energies to figure out. To return to the theme at the outset of this chapter, producing is a way to cultivate and sustain attention, what writer Nadia Asparouhova evocatively calls "putting your brain into... 'developer mode', from which you can write new rules that govern your thoughts and behavior".[10] Now we conclude by considering six macros, each a nudge for moving a track—and ourselves—forward by helping use what we already have in new ways.

One thing at a time. As discussed in the get-each-layer-right section above, finish one musical step before tackling the next. If you're working with a beat, customize each sound in the drum kit before playing a drum part. If you're playing a drum part, record it as a complete phrase before processing with effects. If you're processing a part with effects, dial in one effect before adding another. In other words, approach each production step as a self-contained goal. One thing at a time brings clarity to a workflow as one's multiple steps accumulate and produce composite sounds impossible to achieve any other way.

Morph timbres. If you've already played a part, reuse it (or part of it) somewhere else. Chords moved to a mono lead two octaves lower become bass; bass minus sustain on its notes becomes percussion; percussion through reverb becomes ambiance. Producer Kieran Hebden (aka Four Tet) explains how he'll re-use MIDI to trigger additional synthesizer parts:

> If I've created a melody with MIDI in the piano roll and it's playing a soft synth, it's so tempting to put that in another synth and reverse it, or shorten it, and see how these things can play off each other.
>
> *(Herlihy, 2021)*

Make copies at different rates of speed. If you have a fast-moving part make a slower-moving copy of it. Or double-time a slow part. Stretch and compress tracks to hear how they sound. Making copies at different rates of speed reveals the expressiveness of one's materials, a topic we'll discuss more in Chapter 8.

Move away from the obvious. If a sound or a chord or a beat or an arrangement sounds too predictable, tinker with it until it sounds less so. There's nothing wrong with the obvious, except in music.

Amplify the pivotal. Discern the element in the music most responsible for how it feels. Sometimes this is a single detail—a lingering note, beguiling chord shift, subliminal mix perspective, hazy artifact, hesitating melody, or a striking textural contrast. Notice what you identify as a track's pivotal element, and then devise a macro for amplifying it.

Stay supple. This is the producer's most useful macro. Devise a system for making music playfully and quickly before the onset of overthinking.

Notes

1 Analogously, from sound waves to ocean waves: in Italo Calvino's novel, *Mr. Palomar*, Palomar grapples with describing the experience of watching a single ocean wave. "You cannot observe a wave without bearing in mind the complex features shaping it and the other, equally complex ones that the wave itself originates" (Calvino, 1985: 4).

2 Artist and writer Jenny Odell sees artworks as "training apparatuses for attention" at various resolutions. She writes: "By inviting us to perceive at different scales and tempos than we're used to, they teach not only how to sustain attention but how to move it back and forth between different registers" (Odell, 2019: 108).

3 And likewise, note Eno and Bette Adriaanse in their book, *What Art Does*, "artists are feelings merchants" (Eno and Adriaanse, 2024: 73).

4 We can understand the phenomenology of *writing* about music as the mirror image of the experience of producing music. Writing takes its cues from music's affective life, bringing its sonic concepts into the textual domain. A musical phenomenology is a writing that sings its analysis, generates its lexicon, and blooms its concepts from the sounds it encounters. Writing that follows the paths of melodies and rhythm cycles, feels the textures of sounds, connects the style with the substance. Writing that describes, evokes, conjures, summons, and elicits music's affecting presence. Writing that revels in music's sensual, tactile, transient, and resounding terrain. Writing that takes music's measure without counting. Writing that foregrounds music's negative space, hears its inherent rhythms, tracing what's unsaid. Writing that augments and amplifies, performs, not theorizes music's enchantments. Writing that clears the cache of clichés, catching music's vibrations, alive at many levels at once. Writing that suspends disbelief that music is magic. In sum, music is a model for a writing that's attuned to musical energy, mystery and aura, structure and gesture, grain and dimensionality, surfaces and depths, shape-shifts and metamorphoses.

5 A decade later, the philosopher Thomas Clifton proposed a similar understanding of music: "Before becoming a cultural artifact, a style, or an object of study, music is a presence" (Clifton, 1983: 80).

6 Writing phenomenologically about the ways of the peregrine, J.A. Baker observed how "the peregrine sees and remembers patterns we do not know exist: He finds his way across the land by a succession of remembered symmetries" (Baker, 1967: 48).

7 Producer Chris Clark (aka Clark) suggests five layers: "The rule of five—that's quite good. The rule of five—five things in a track" (Nicholas, 2014).

8 Listen, for example, to Autechre's "Bronchusevenmx".

9 For example: the violin and cello were invented in the early 1500s, the piano was invented around 1700, and the first piece for these instruments, Joseph Haydn's Piano Trio No. 1 in G Major, was written in 1795.

10 Asparouhova came to this realization while teaching herself *jhanas* meditation, whose perceptual insights she describes in her essay *Does meditation experience improve success with the jhanas?* (Asparouhova, 2024).

4

MINIMALISMS AND CONSTRAINTS

Introduction

In a 1997 *Sound On Sound* interview, the composer and keyboardist Harold Budd explained how he used minimalism and constraints to shape his creative process. Weary of anything in the music producer's studio that could distract him, Budd was fond of imposing arbitrary restrictions that kept his focus as narrow as possible. It was restrictions that helped him find the "musical sense" in small things with which to make something exquisite:

> I like the arbitrary restrictions that one places on oneself, so that you don't get scatterbrained and don't reach for everything that's available. Keep your focus very narrow: just this and nothing more, and make that absolutely exquisite and don't get sucked into distractions, don't listen to the siren chorus singing across the waves about this keyboard that has a billion sounds in it. I couldn't care less about things like that. They just get in the way. I'm not bragging, but the way I work is that I focus entirely on a small thing and try to milk that for all it's worth, to find everything in it that makes musical sense. A studio gives you the freedom to do that, but it also gives you the freedom to do everything, and to me everything is a tyranny.
>
> *(Tingen, 1997)*[1]

Budd's strategies are ever relevant today, especially for producers who feel overwhelmed by the freedom to do everything in the DAW, suffering from what the writer Alvin Toffler presciently called *overchoice*.[2] But we can use minimalisms and constraints to set boundaries on our music making to sidestep overchoice and evade the siren chorus. It's easier to make music when we limit our field of search.

This chapter explores facets of minimalism and constraints as they intersect with music production. We examine the minimalist mindset, minimalist approaches to production, in-the-moment constraints, process not outcome, the elegance of economy, and demo thinking. Also, because its topic is minimalism, the chapter is brief.

DOI: 10.4324/9781003585145-5

FIGURE 4.1 The minimalist studio of Colleen (photo by Cécile Schott)

The minimalist mindset

More important than having a billion sounds to choose from is adopting a minimalist mindset so that one good sound is enough. A minimalist producer hears less as more. She simplifies and clarifies music production, framing the experience by reducing the number of variables and options to consider while building a track.[3]

The producer Cécile Schott (aka Colleen) makes music using just one or two analog synthesizers and a few effects pedals, forgoing VST instruments altogether and using her computer only as a recording device. Working with the Moog Grandmother, a monophonic synthesizer with a small keyboard (Figure 4.1, upper right), Schott explores different ways of playing the instrument. "There is a massive difference between playing a synth by hand, using the arpeggiator, a quantized sequence, or an unquantized sequence, and I used all four techniques" she explains (Obenour, 2023). Schott evinces a minimalist mindset when she notes that even though her recordings, such as *Le jour et la nuit du réel* (2023), are "voluntarily restricted in terms of gear", their sound does not "feel restrictive at all in terms of musical and sound possibilities" (Moller, 2021).

We can learn from Schott's voluntary gear restrictions. The minimalist might begin with a single sound—like the oscillator Sarah Davachi recommended listening to in Chapter 3—whose quality is the focal point, fascinating in and of itself, a musical end, not a means. The minimalist hears more as less because more strays from a single sound's clarity. More multiplies and meanders, while less is tensioned with potential. The minimalist constructs music while constraining the elements that constitute it:

a minimum of sounds,
structure,
effects,
and compositional intervention

lead to

a clarity of instrumentation,
direction,
texture,
and feeling.

Adopting a minimalist approach helps producers achieve focus and purpose in their work.

Minimal approaches to music production

It's not just Harold Budd and Cécile Schott who notice the usefulness of a minimalist workflow. Kraftwerk percussionist and producer Karl Bartos makes the case that the *minimalist concept* is inherent in electronic music. For Bartos, this aesthetic emerged from the capabilities of early (circa 1970s-90s) electronic musical technologies. The sound of looping sequencers and drum machine beats evoked the trance aspects of enduring non-Western musical traditions. In fact, these technologies' capabilities to effortlessly repeat patterns introduced a mechanized analogue of the hypnotic qualities of non-Western musics, such as West African drumming and Indonesian gamelan. "The sequencer's automatic sound sequences and the drum machine's loops hypnotized me" Bartos recalls. "These black boxes brought the trance quality of African, Indian and Asian musical cultures into pop music, a quality that had been the starting point for the minimalist concept" (Bartos, 2022: 353). In this way, minimalist thinking is deeply rooted in electronic music production.

Many producers, working across a range of styles, have adopted a minimalist approach to production. Let's consider four examples.

Producer Will Long (aka Celer) makes music in a single gesture with just a few sounds. He uses his DAW, Logic Pro, as a digital tape recorder for arranging acoustic instrument parts (guitars and keyboards) and field recordings:

> Usually since I'm using *a limited amount of equipment with few options*, the sound direction stays largely focused in the same direction, and strays from there. I don't try to build pieces too often that have too many different types of sounds, or things that were made at different times overlapping simultaneously—usually what is together was all made around the same time, in one sitting.
>
> *(Headphone Commute, 2014, italics added)*

Producers' minimalism often arises from the limitations of their musical tools. Writer and Art Of Noise conceptualist Paul Morley recalls how, in the 1980s, hardware samplers shaped the group's sound. Technologies of the time such as The Fairlight CMI could only store a few seconds of audio. This limitation forced Art Of Noise to use the sampler resourcefully. They had to figure out how to achieve maximum effects from the Fairlight's minimal sample length:

> What created part of the original sound of The Art of Noise were the limitations of the technology. The early sampler, The Fairlight, had like a 2 second sampling window and you put in a nice bright sound it and it came back all dull and horrible. That forced you to do interesting and creative things with it. It had a certain sound, and if you listen to those early Art Of Noise albums, they have a certain sound and that is determined by the limitations of the technology. Once the technology ceases to have limits, it's almost like having too much freedom. It's difficult to work, really.
>
> *(Gourley, 1999)*

Producer Simon Green (aka Bonobo) similarly recalls how early in his career, circa 2000, he worked with Akai hardware samplers to make his recording *Animal Magic*. The Akai S900's relatively limited sampling time and polyphony were constraints that pushed Green to devise a minimalist strategy as to which sounds to use and how to use them. Green still has this *make every sound count* mindset today:

> We're in the digital era where you have infinite boundaries of what you can do. But what I liked about that era of *Animal Magic* was working within the confines of these samplers that only had 32 seconds of sample memory. So everything in the track had to really count. Nothing could be disposable.
>
> You had to make that 32 seconds of sample memory count and there's only 8-note polyphony, so you can only have eight sounds at once. And I think that was a useful parameter to work in early on, because it just makes everything a bit more considered...Every sound has to contribute something to the track and not just kind of pile those [sounds] in there for the sake of it. And the idea of things are not going to get better if you just throw more sound at them. *It was an exercise in minimalism*, which I think is something I've tried to take with me as much as I can.
>
> *(Tape Notes, 2024, italics added)*

Autechre member Sean Booth explains the minimalist approach he and Rob Brown took in the early days of their career when they tinkered with hardware keyboards, samplers, and drum machines to explore these machines' potentials for making music. Autechre's early minimalism entailed *using a single instrument to make every sound in a track*. This helped Booth and Brown learn what an instrument could do:

> I used to have a thing where if I had a keyboard or drum machine, I'd try to make tracks only using that. Even though it was a bit shit and I didn't know what I was doing, it forced me to learn every single thing it could do because I needed to use it for every single part of the track. That'd open it up for me.
>
> *(Muggs, 2016)*

Booth emphasizes that Autechre's strategy of working with a single machine and learning every single thing it could do had an ambitious creative goal, which was "just get the little thing and fucking max it out. You know, get it doing shit that other people aren't doing" (Moreno, 2024). Like Bonobo, Autechre's early minimalism continues to inform a sophisticated musical practice. What was once a minimalist mindset born of trying to understand how a technology works has become an ideal headspace. Reflecting on his present-day programming in a Max-based musical system, Booth explains that "I'm quite often trying to access that space that I was in before I knew how any technology worked, when I was hearing records and imagining what the sounds might be. That to me is a really valuable thing" (Muggs, 2016).

In the moment production constraints

An elegant form of minimalism is imposing constraints on our production practice. A constraint is that which limits us in some way—something we can push against or work around, a situation or configuration or boundary that requires us to be resourceful and focus on what we have instead of what we don't. As shown in the examples above, common artistic constraints are limited materials, limited tools, and limiting conventions of practice. Producer Rafael Anton Irisarri suggests taking a proactive approach to turn one's limitations into assets. "The key" he explains

> is to identify your limitations and use them as parameters to develop your unique aesthetic. What might seem like a flaw can actually become a valuable asset in your praxis. Instead of thinking, 'If I had X piece of gear, I could do so much better music,' focus on what you can create with the tools you already have, not what you wish you had.
>
> *(Truss, 2024a)*

Whether self-imposed or inescapable realities of the project at hand, constraints help artists make progress by reducing their options.

A serendipitous kind of production constraint is one that arises organically in a workflow moment. In-the-moment constraints are not premeditated at the outset of a project (e.g. "My track will *only* use *these four* sounds...") but instead appear and evolve as we learn which workflow moves are yielding the coolest-sounding results, and we adapt accordingly. This illustrates architect Christopher Alexander's observation that nature's alive-at-many-levels-at-once quality, the topic of Chapter 2, is achieved *dynamically* through feedback loop adaptations:

> Things that are good have a certain kind of structure. You can't get that structure except dynamically. Period. In nature you've got continuous very-small-feedback-loop adaptation going on, which is why things get to be harmonious. That's why they have the qualities that we value. If it wasn't for the time dimension, it wouldn't happen.
>
> *(Brand, 1995: 21)*

In the moment constraints arise out of our feedback loop adaptations to our musical system as we interact with it. Like the mechanic in Robert M. Pirsig's *Zen and the Art of Motorcycle Maintenance*, the attuned producer takes cues from the affordances of his musical machine. "He isn't following any set of written instructions" says Pirsig, "because the nature of the material at hand determines his thoughts and motions, which simultaneously change the nature of the material at hand" (Pirsig, 1974: 168–69). Producer Martin Stimming (aka Stimming) explains how the musical material at hand offers clues as to what a track needs to reach its finished state. In some cases, it's the producer's mindset that's in need of shifting—in other words, we need to adapt to the music. It's at this point, says Stimming,

> where it's not so much about what you want with the track anymore, but what the track wants you to do with it. To figure out this change of perspective is the only way to successfully finish a track. Of course it's intuition, but at a certain point the music is the boss. And you always recognize it when you want to finish a very fancy idea for example and it doesn't work. You build the arrangement and it doesn't work, you try something else and it doesn't work. If you just listen to what the track wants, it's much easier.
>
> *(Stimming, 2016)*

In sum, in addition to imposing production constraints from the outset, the producer can cultivate constraints as needed in the moment, as a way of listening to what the track wants. Cultivating constraints is a very small feedback loop adaptation, a dynamic way of interacting with our music in progress.

Process, not outcome

The saying, *process, not outcome* is the most actionable minimalist advice for doing artistic work. We can't control our work's outcome—whether it will succeed in doing what we hoped it would do, whether others will find it beautiful or useful, and so on. But we can choose and commit to a process that becomes a workflow. A process often seems arbitrary until you commit to where it leads, at which point it blooms, revealing its potentials. Consider artists who've committed to a process: Roy Lichtenstein using Benday dots to animate his paintings, Matsuo Bashō writing Haikus using 17 syllables in a 5–7-5 pattern, Steve Reich replicating the sound of going-out-of-sync tape recorders with live percussionists.[4] The producer discovers dynamism by focusing on musical processes instead of musical outcomes. Process is both a way of thinking minimally and a powerful production constraint. Two further examples illustrate the point.

Production as diary: Michiru Aoyama's daily ambient releases

Since 2021, ambient music producer Michiru Aoyama has been recording and releasing an 8- to 12-track album *each day*. Aoyama's workflow, or what he calls his "compositional mould" is playing electric guitar or synthesizer (a Sequential Prophet-10) through a system of effects processing. The workflow is both minimalist and constrained in that neither the instruments nor the processing effects change from day to day. Instead, Aoyama's self-imposed creative goal is to continually find new ways of making music with a limited set of tools. A tacit constraint that helps him realize the goal is suspending judgment about the quality of his musical system's output. Rather than worrying about how good or bad the day's music is or laboriously editing it to perfection, Aoyama releases exactly what he generates in a morning's session. "Writing a song is like keeping a diary" he says. "It is a trial and error process, but I release it all, the good sound quality and the bad sound quality without any concealment" (Fitzpatrick, 2023).

Acoustic re-imaginings: Taylor Deupree's "Stil."

Another example of production process over production outcome is producer Taylor Deupree's *Stil.*, a 2024 reimagining of his 1998 recording of the same name. Deupree composed the original out of short loops woven into tapestries of pulsing ambiance. Deupree's work was part of the *microsound* music scene of the late 1990s and early 2000s. Much of this experimentally-oriented scene focused on using samplers and granular synthesis techniques to make music out of tiny sound particles (cf. Roads, 2004). In his article *The Aesthetics of Failure*, composer Kim Cascone summed up the scene with the term "post-digital" and explained how "specific [musical] tools themselves have become the message" (Cascone, 2002). Deupree recalls that many microsound musicians pursued process-based music production as a constraint for using the computer as a musical instrument:

> I think my most direct relationship with this type of [process-based] approach was in the early 90s in the birth of the 'microsound' movement when we were first using the computer itself as a musical instrument. A lot of the work we were doing was seeing where we could push the boundaries to or how far we could deconstruct a sound. It was, at its heart, an entire genre based on process. That isn't to say there wasn't a lot of lasting and interesting music made in those days, there was, but a lot of it was rigid and formal. But those experiments had to happen and led electronic music to where it is now where the computer, still used as an instrument, has taken on a much more mature and varied sound.
>
> *(Sannicandro, 2017)*

In 2024 Deupree revisited *Stil.* through what he calls an "acoustic re-imagining" by translating the recording's electronically-produced sounds into acoustically-produced ones. With the help of composer Joseph Branciforte, who transcribed and arranged the original's asynchronous loops into notated instrumental parts, Deupree placed constraints on his re-imagining. First, all the sounds would be performed live using orchestral instruments (clarinet, vibraphone, cello, double bass, flute, and percussion). This would ensure capturing some degree of the performers' touch, a topic we explored in Chapter 2. "We wanted to do it acoustically" Deupree explains, "and just simply having the player play a loop over and over, it's going to change naturally" (Hanging out with Audiophiles, 2024). Second, since the sounds on the original recording were repeated for long stretches, Deupree and Branciforte added written directions into the musicians' parts. These texts were analogous to the producer's automation curves that bring life to a track. "Writing in dynamics, writing in phrasings, writing in accents, writing in little commentary. It's like something to engage their minds. So there's this whole other dimension to [the music] where it's not a synthesizer" (Hanging out with Audiophiles, 2024).

The final constraint on *Stil.* is that it received minimal post-performance editing and processing. Instead of altering the recording with effects and volume automations in the DAW, Deupree specified diminuendos in the score so that "the players would play more quietly over the course of 30 seconds or a minute or whatever the fade time was. And when you have a clarinetist or a cellist trying to play quieter, it's not that easy." In sum, through constraints Deupree brought human touch and process into his re-imagining; *Stil.*'s acoustically realized fade-outs create pleasing sonic artifacts without the need for electronic effects. "It has that degradation" he says, "kind of almost that tape-like degradation [with] the tactility and the noise of the bow" (Hanging out with Audiophiles, 2024).

The elegance of economy

In music production there are myriad ways to get a sound from here to there. Let's say you want to thicken a sound: Do you double it with another one? Process it with an effect? Or alter it from its inside out by adding or changing oscillators? If you want to make a beat: Do you play it in or program it? Use a loop or manipulate a sample? If you want to mangle a sound into something entirely unrecognizable, how do you do this?

There are many routes to usher a sound from here to there, but far fewer routes to achieving this in an elegantly economical way. Producer Brian Eno (elegantly) defines production elegance as using simple methods to get elaborate results: "I really admire economy more than anything else: elegant ways of making big things happen—which is the opposite of what normally happens in a studio, where you have clumsy ways of making small things happen," he says (Ward, 2021).

The producer seeks to move a sound from here to there in the fewest possible steps, not because it's the easiest way forward, but because it's the most elegant. As Eno says, elegance is economy, and economy in production entails finding *elegant ways of making big things happen*. Elegance, in other words, is doing the minimum to achieve maximum results. To return to our examples above, we could thicken one sound with any other one, but that could involve long periods of searching for, and auditioning random sounds without a relationship to the original. A more elegant way is to copy the sound, alter the copy, and then use this as a thickening agent. If your original sound was say, a metallic keyboard tone, you could make its copy a warmer hue that maintains a connection to it.

If you're making a beat, using a loop is easy, but maybe not elegant. Instead, play your own beat imperfectly, which will generate interesting inconsistencies to work with. Or if there were artifacts in your original metallic keyboard tone, you could resample these to make a palette of unusual drum samples. The producer leverages the quirks of a performance, knowing that resampling artifacts is a more elegant process than dragging in a drum loop and hoping it fits. Even if someone else's loop fits, it has *zero relationship to our work*. That's not elegant.

Mangling sounds can of course bring us down rabbit holes of trying out effects and effect chain processing, but a more elegant strategy is to reuse only those effects already in the track. This is an economical way of reducing choices to just a few to *limit your field of search*. Instead of pondering the overwhelming possibilities of *I could try anything on this sound*, explore alternate uses of existing effects. In our example, the effects used to thicken the metallic tone could also process the tone's artifacts into unusual drum sounds. Producer Astrid Sonne amplifies this idea:

> Rather than working with sound and timbre in order for them to meet production ideas, I try to *bring out different musical parameters from the material I'm already*

> *working with.* I like to look at the sound as a small sculpture I can examine from different angles and put together the bits and parts I like the most.
>
> *(Fischer, 2021, italics added)*

Limited options allow the producer to work unimpeded. They inspire confidence that a few paths are enough to generate new material via making small changes to sounds and devices already in play elsewhere in the track. In sum, the producer thoughtfully re-uses what the music already has, working with the constraints of what is close at hand to turn a minimum into a maximum and generate *elegant ways of making big things happen.*

Demo thinking

There's no shortage of ways to begin a track, because we can begin anywhere and build from there. But once we've begun, we can feel a self-imposed pressure to make something big happen, to prematurely optimize what we have, or to turn a sketch into a finished track. One way to reduce this pressure is to treat tracks-in-progress, for as long as possible, as demos.

Demos have a long history in music production. In the 1970s and 80s demos were song sketches one recorded on cassette to give others a sense of how the music goes.[5] Usually a demo would be re-recorded before release, but occasionally it was released as it was. Bruce Springsteen famously released the demo versions of the songs for *Nebraska* once he realized that it was impossible to recreate their feeling on "proper" studio versions recorded with his band. Referring to his Teac TASCAM 144 four-track cassette recorder, Springsteen recalls how he and his engineer, Toby Scott, "were gifted by our gear and the equipment that we were casually using at that moment" (Zanes, 2023: 157). According to Scott, Springsteen said, "there's just something about the atmosphere on this tape. Can't we just master off this?" (Keller, 2007). What makes a demo's unpolished atmosphere powerful is the *demo thinking* it embodies. Springsteen hints at this idea when he talks about *casually using* his Teac recorder. We can imagine him just hitting Record and playing through the songs in a single flowing gesture. Easy.

When I'm making what I consider a demo, my aim is to capture the essence of an idea, buoyed by knowing that I can always improve it later. I'll quickly record a few parts, and rather than obsessing over the right sounds I settle on the first ones that get the job done. This keeps the process loose, "making it fun with as low stakes as possible" as producer Harley Streten (aka Flume) puts it (Willings, 2022). Or as producer Chris Clarke (aka Clark) explains, thinking in demo terms keeps oneself "in this zone of just dabbling", in the flow of "getting stuff down really quickly and rendering to groups and just condensing mass of recording and then letting stuff fall away that isn't hitting me" (RA Exchange, 2022). With a few parts recorded, I might quickly arrange them to hear how they work together, and if some effects are needed, I'll add these as I go, using whatever is close at hand. A demo can be finished in a single session. We stop when we've run out of ideas for what to do next, or when the track doesn't seem to need anything else, at least for now.[6]

In sum, the value of demo thinking is that it prevents producers from getting ahead of themselves and obsessing over ever more refined versions of the music. This idea connects to the themes of minimalisms and constraints. Just as we limit our field of search, so too can we limit our ambitions: although we intend to improve it later, in the meantime the music may be refined enough. Demo-making keeps us in the moment because it's casual, exploratory, and in the zone of just dabbling. The music feels unfinished, the stakes low. And this is precisely how a demo exerts its magic. A demo compels us to work quickly and uninhibited, because we don't know if we'll ever return to this track-in-progress. Demo thinking is a necessary fiction to get music going, and sometimes, even done.

Notes

1 In a later interview Budd humorously described himself as "I'm so minimal, I'm not even minimalist" (Nelson, 2014).
2 "People of the future may suffer not from an absence of choice but from a paralyzing surfeit of it. They may turn out to be victims of that peculiarly super-industrial dilemma: overchoice" (Toffler, 1970: 278).
3 Consider a culinary analogy: minimalism, explains the chef Ferran Adrià, is "the method by which maximum 'magic', or sensory appeal, can be created with minimum ingredients" (Adrià, 2010: 317).
4 In his essay "Music as a Gradual Process", Reich described musical processes as the musical thing itself: process gives "one a direct contact with the impersonal and also a kind of complete control…By 'a kind' of complete control I mean that by running this material through the process I completely control all that results, but also that I accept all that results without changes" (Cox and Warner, 2004: 305).
5 For example, listen to Michael Jackson's 1981 demo for "Billy Jean" (Jackson, 1981).
6 I share track demos on the Atelier page of my blog, brettworks.com.

INTERLUDE

AI in Electronic Music Production

Introduction: souls trapped in circuitry

In early 2024 an AI company called Suno posted on Soundcloud an excerpt of an original song called "Soul of the Machine" to show off its generative AI text-to-audio technology (Rolling Stone, 2024). The song was a collaboration between Suno's AI model, which generated the music, and ChatGPT, OpenAI's chatbot, which generated the song's lyrics and title. Based on the text prompt *solo acoustic Mississippi Delta blues about a sad AI*, "Soul of the Machine" conjured a blues singer accompanying himself on acoustic guitar. The voice plaintively sings of the loneliness of being trapped inside AI, as felt from the computer's perspective: "I'm just a soul trapped in its circuity/My heart used to be but now it's cold and weak…" Suno's AI wasn't deep-faking the voice of any specific blues singer. Instead, it had generated a stylistically accurate facsimile of a generic one whose singing and guitar playing sounded believable—like a real song. What's striking about the excerpt is that the sound and inflections of its AI bluesman are convincing as a *performance*. If there were a music listening Turing test, "Soul of the Machine" would pass it.

Increasingly believable AI-generated songs are just one element of a larger, evolving story of AI's incursion into music production, and the pace of progress in recent years has been hurried. There are now numerous AI tools for generating audio samples or entire tracks, deep faking the voices of pop stars, and automating various aspects of music's production—from songwriting to chord generating, sound designing to mastering. With each new AI tool that appears (seemingly every few weeks), a corresponding cheerleading or alarmist AI versus music-themed article pops up in industry magazines. "For many artists, AI is not a threat but an opportunity" one article in MusicTech notes optimistically and rather uncritically. "A new instrument that could usher in a musical renaissance comparable to the advent of 1980s sampling culture" (McEvoy, 2022). Time will tell if our era is a renaissance or a time of decline as musicians find themselves amidst an excess of AI tools trained on the human skills they're intended to replace and wondering what it all means. For the

DOI: 10.4324/9781003585145-6

moment, AI is a monkey wrench in music production's conventional ways of creation. The uses of AI pose questions about the nature of producers' work, the sources of their ideas, and even the notion of artistic originality. This interlude takes measure of some prominent AI-in-music production developments and their impact. We begin with a history of proto-AIs in music, then move to contemporary practices to consider the implications of prompt-based music production, three case studies of producers collaborating with AI models, and the rise of vocal cloning/licensing. The conclusion extracts production lessons from AI processes and concepts: the importance of taste, artistic influence as dataset, implementing musical prompts, and understanding musical creativity as error artifact.

A history of proto-AIs in music

Today's AI music technologies have predecessors in the history of western music. Long before digital computers, composers experimented with algorithmic-based techniques, which can be considered as a kind of proto-AI. These techniques constituted a non-human intelligence that helped composers by automating or generating options for various aspects of the creative process. An *algorithm* is a set of instructions for solving a problem in a limited number of steps. A popular algorithm developed and widely used in late medieval and renaissance times was the *canon*, a musical form in which multiple singers or instrumentalists share the same melodic line but begin it at different times.[1] An example of a canon is the French nursery rhyme, "Frère Jacques". In terms of performance practice, it was common for a composer to write out a single melodic line and give the performers instructions—the algorithm—that the line be sung or played as a canon. In essence, canons are automated counterpoint generators.

Other proto-AI techniques appear over western music history in the form of randomness generators. For example, in the 18th century, composers sometimes used dice to randomly produce music upon a collection of precomposed fragments. In 1787, Mozart allegedly wrote *Musikalisches Würfelspiel* ("musical dice game"), a piece intended as a parlor game or system for constructing music out of 176 one-bar fragments (Figure I.1). In the early 20th century, Arnold Schoenberg invented a compositional system known as *serialism*, a form of the *twelve-tone* technique that sought to comprehensively control music's pitch, rhythm, and timbral parameters according to "pre-composed 'series' and 'matrices' of values" (Maurer, 1999). Serialism's methods were the opposite of random, but what made them proto-AI was their outsourcing of a composer's expression to numerical rules for controlling musical parameters. "The serial method in effect 'automated' many of these parameters" notes John A. Maurer, "by determining the order in which each must occur in a piece" (Maurer, 1999). In other words, serialism was a composing algorithm.

In the 1950s, John Cage outsourced and automated his creativity by using the Chinese *I-Ching* (Book of Changes) to generate chance procedures by which to make compositional decisions and model the workings of what James Bridle calls the "more-than-human-world" (Bridle, 2022: 234). One of the first works Cage used the *I-Ching* to compose was *Imaginary Landscape No. 4* (1951). The work, a kind of early electronic soundscape exploration, is scored for 24 performers who "play" twelve radios as

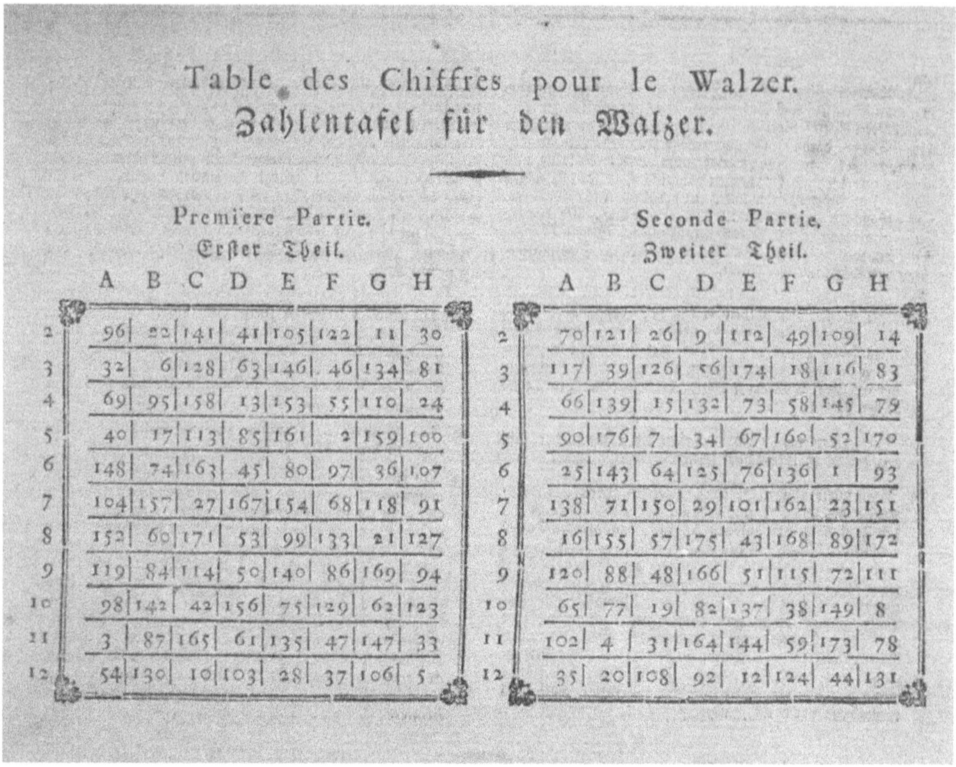

Table des Chiffres pour le Walzer.
Zahlentafel für den Walzer.

Premiere Partie.
Erster Theil.

	A	B	C	D	E	F	G	H
2	96	22	141	41	105	122	11	30
3	32	6	128	63	146	46	134	81
4	69	95	158	13	153	55	110	24
5	40	17	113	85	161	2	159	100
6	148	74	163	45	80	97	36	107
7	104	157	27	167	154	68	118	91
8	152	60	171	53	99	133	21	127
9	119	84	114	50	140	86	169	94
10	98	142	42	156	75	129	62	123
11	3	87	165	61	135	47	147	33
12	54	130	10	103	28	37	106	5

Seconde Partie.
Zweiter Theil.

	A	B	C	D	E	F	G	H
2	70	121	26	9	112	49	109	14
3	117	39	126	56	174	18	116	83
4	66	139	15	132	73	58	145	79
5	90	176	7	34	67	160	52	170
6	25	143	64	125	76	136	1	93
7	138	71	150	29	101	162	23	151
8	16	155	57	175	43	168	89	172
9	120	88	48	166	51	115	72	111
10	65	77	19	82	137	38	149	8
11	102	4	31	164	144	59	173	78
12	35	20	108	92	12	124	44	131

FIGURE I.1 Manuscript for *Musikalisches Würfelspiel* ("musical dice game"), attributed to Mozart, 1792

instruments: one player controls the radio's station dial and the other controls its volume. In 1958, Lejaren Hiller and Leonard Isaacson created *Illiac Suite*, a four-movement string quartet considered the first score composed by computer. Hiller and Isaacson programmed instructions into the ILLIAC, a vacuum tube-based, 4000-pound mainframe behemoth at the University of Illinois, to randomly generate musical materials. At the outset of *Illiac Suite*'s 1960 recording, a narrator explains how the project was

> a research record...to determine whether an electronic digital computer could be used to compose music...The computer was programmed to select musical notes randomly, subject only to general instructions derived from rules of musical composition which were stored in the computer. It printed out sequences of letters and numbers which were then translated into conventional music score and played by human musicians.
>
> *(Bell Telephone Laboratories, Inc., 1960)*

In the early 1960s, Iannis Xenakis created a more sophisticated computer program, Free Stochastic Music, to calculate randomly determined or "stochastic" probability distributions to assist his composing.[2] As if anticipating today's AI text-to-audio sample pack generators, the output of this software was "not the composition itself but material with which Xenakis could compose" (Maurer, 1999). Xenakis imagined each piece of music as a "mental machine" and presciently foresaw our era's "utilization of electronic computers in musical composition" (Xenakis, 1992: 132–133).

In 1981, when the first personal computers such as the Apple II appeared, David Cope began a decades-long pursuit of programming software to help him compose. Guided by a belief that creativity is fundamentally recombinant, Cope's first software was Experiments in Musical Intelligence (EMI), or Emmy. What made Emmy different from the proto-AI randomness excursions of Cope's predecessors is that the software could generate its own grammar for making music. This was possible because Cope built the software upon a database of rules that describe different compositional strategies, such as voice-leading in the style of Baroque era four-part chorales (Wilson, 2010). With this musical metadata as its input, Emmy could output stylistically convincing music in the style of Bach, Mozart, and even Cope himself. Cope trained Emmy on Bach's chorales so that the software could analyze their voice-leading and counterpoint and use those data "to create new music in the style from the music in the database, but not actually replicating any actual piece in the database" (Heater, 2013). In 2010, Cope predicted that "in the next 10 years what I call algorithmic music will be a mainstay of our lives" (Adams, 2010). But, as of this Interlude's writing, that has yet to happen.

A few years before Cope began writing programs to compose in the style of Bach, Brian Eno had figured out less technical generative methods by which to make music. As we learned in Chapter 2, Eno's method was inspired by ideas in cybernetics, specifically the idea that a system's dynamics, rather than the composer's intentions, can be a source of artistic ideas. Eno sought ways of setting up a musical system and then riding on the dynamics of the system—like a wave—to grow music. He called this system *generative*. Perhaps the idea was in the air, because Eno found confirmation for his approach in the 1960s tape music pieces of Steve Reich.[3] Reich had used samples of voices to make "It's Gonna Rain" and "Come Out", looping the voices on two tape machines so that they played in unison. But due to the machines' minute speed difference, the samples serendipitously went out of sync and created ever-changing rhythmic echo effects. Reich noticed these echoes and developed his *phasing* canonic technique on them.

Eno was struck by the powerful, shifting perceptual aspect of Reich's tape music, specifically the way "It's Gonna Rain"'s out-of-sync composite sounded more than the sum of its parts. The source of the music's power was a sonic moire effect or interference pattern produced when two templates that are similar but slightly offset are overlayed. Taking, for example, "two identical ruled transparent sheets of plastic, superposing them, and rotating one about its center as the other is held fixed" (Weisstein, n.d.). Eno realized that Reich's tape music processes were a generative system with which to compose:

> Now a moire pattern... is actually a very good analog of the Steve Reich piece ["It's Gonna Rain"] in action. Something happens because of one's perception rather than because of anything physically happening to these two sheets of plastic which produce an effect that you simply couldn't have expected or predicted. I was so impressed by this as a way of composing that I made many, many pieces of music using more complex variations of that...All of my ambient music I should say, really was based on that kind of principle, on the idea that it's possible to *think of a system or a set of rules which once set in motion will create music for you.*
>
> *(Eno, 1996, italics added)*

What these examples—Cage's use of the I-Ching, Hiller and Isaacson's use of ILLIAC computer, Xenakis' and Cope's softwares, and the generative musics of Reich and Eno—have in common is that they rely on randomness, however generated, as a kind of alien intelligence with which to compose or by which to enable a composing process to unfold. But as Marcus du Sautoy reminds us, randomness isn't inherently creative, and an alien intelligence still requires a human to assess and curate its artistic results. "Creating art" he says, "is not a problem-solving activity" (du Sautoy, 2019).

Contemporary musical AI practices I: AI text-to-audio developments

In the early days of AI models such as ChatGPT, DALL-E 2, MidJourney, and Stable Diffusion, there's been much excitement around the technologies' capacity to generate novel texts, images, and musics around any topic or theme with a believably human level of sophistication and complexity. Built on deep learning algorithms, these AI models have been described as "clairvoyants", which compress knowledge "of the past into the present in a new arrangement" (Carr, 2025: 181). This clairvoyant generating is possible because the softwares use large language models (LLMs) trained on billions of examples culled from the internet. As Stephen Wolfram explains, the LLM ChatGPT works by trying "to produce a 'reasonable continuation' of whatever text it's got so far, where by 'reasonable' we mean 'what one might expect someone to write after seeing what people have written on billions of webpages', etc." (Wolfram, 2023: 1).

A key to interacting with AI is the text *prompt,* which points it in a general thematic direction. A prompt takes the form of a question or a directive, and popular prompts include famous author or artist names, cultural references, places and eras, and fantasy scenarios. Nuanced and unconventional prompts can generate nuanced and unconventional results. For example, an entrant in a 2022 digital art competition used a prompt on the theme of "space opera theater" to generate a richly detailed, first prize-winning work, *Théâtre d'Opéra Spatial* (Roose, 2022). Within a prompt there are ways to further constrain an AI's search. A common constraint is to use the qualifier *in the style of* as a way of linking two disparate ideas. Upon the release of Chat GPT-4 in 2023, a much-discussed humorous prompt example was one asking for "instructions for removing a peanut-butter sandwich from a VCR, written in the style of the King James Bible" (Newport, 2023). While LLMs' prompt-constrained results aren't always aesthetically thrilling, they can be uncannily realistic. In part this is because implicit in AI technologies is a conception of creativity as a fundamentally imitative and derivative process. To use Wolfram's phrase, creativity is a *reasonable continuation* of something that already exists. Indeed, the most talked about AI-generated cultural artifacts seem to demonstrate that in order for AI models to pass the Turing Test of reasonable, human-like creativity, they need to be able to *generate something in the style of something else.*[4] Like, for example, Suno's bluesman singing a blues about a sad AI.

A significant AI development over the past five years was the appearance of text-to-audio and text-to-music tools. In 2020, OpenAI released Jukebox, neural net-based software for generating music "in a wide variety of genres and artist styles" (OpenAI, 2020). In 2022, Riffusion released their self-same tool, based on their text-to-image generator Stable Diffusion, but using images of sound (spectrograms) rather than audio to generate music (Coldeway, 2022). In 2023 Google developed several AI music tools: Lyria, Dream

Track, and Google MusicLM. Lyria, a model made in partnership with YouTube, invited users to "simply enter a topic and choose an artist from the carousel to generate a 30 second soundtrack", whose music, voice, and lyrics are in the style of a selected artist (Google DeepMind, 2023). Dream Track and Google Music LM are software for creating original songs in the style of famous singers (Veltman, 2023) and audio files of instrumentals from written descriptions (Yim, 2023). By 2024 Suno, Udio, Meta, and Stability. ai had each released text prompt-based tools for generating audio and music. For example, Stability.ai's Stable Audio 2.0 can "produce melodies, backing tracks, stems, and sound effects, thus enhancing the creative process" (stability.ai, 2024). The company claimed its AI "can generate songs up to three minutes in length, complete with structured compositions that include an intro, development, and outro, as well as stereo sound effects". On its website are examples of unimaginative musical prompts, such as "calm meditation music to play in a spa lobby" and "lofi hip hop beat, chillhop".

Also in 2024, the music technology company Output, known for its innovative effect plug-ins such as Portal and Thermal, launched the Beta version of Co-Producer, a web-based "set of AI capabilities", intended to help producers "unlock more time creating and less time searching." Output's first Co-Producer product, Pack Generator, was an AI music generator of royalty-free tracks, stems, and sample packs based on text prompts. "Kick start new ideas with a simple text prompt" the website suggested, illustrating with another bland example: *Driving through LA at night, synth pop 91 bpm.* Output framed the value of its AI as a labor-saver: "stop digging for samples, just describe them" (Output, 2024). This author's attempts at using Pack Generator generated mixed results. My prompt "metallic drum samples" generated heavy metal-style instrumentals, "ocean sounds" generated ambient chords, and "marimba chords" generated synthesized African kalimba sounds. None of these sounds were compelling.

Let's pause for a moment to consider two points about using prompt-based AI tools to generate samples, or even entire songs. The first point is that such tools impinge upon the craft of making music and the producer's quest to discover serendipities and create sounds alive at many levels at once. While there is admittedly *some* artistry involved in devising just the right prompt, is prompt-making merely a way around having to develop production competency? A second point is whether using prompt-based AI tools is even necessary, because, as this book has shown, there's already a plethora of sound sources in one's DAW and previous work (e.g. sample material, effect chains) to draw on. Surely devising one's own workflows for generating novel sounds is preferable to using AI, if only because one's own labor is more exciting, personal, and meaningful? Of course, as prompt-based models improve, prompt strategies may enter into many a producer's workflow.

This brings us to two cases of prompts in music production integration. In 2025, the subscription-based sample library Splice incorporated its entire catalog into PreSonus' DAW, Studio One. For thirteen dollars a month producers can have Splice inside this DAW, facilitating searching for and dragging-dropping royalty-free samples into projects. Similarly, Output's Co-Producer became a DAW plug-in (Figure I.2) that listens to a track's output "and finds audio that fits your session" (Output, 2025). Producers can refine the AI-curated, musician-made sample results with key suggestions and text prompts. But in the comments to the software's walkthrough video on YouTube, a perceptive musician observed that a problem with Co-Producer is not just that "the samples sound pretty basic." The plug-in also bypasses producing's core cognitive component.

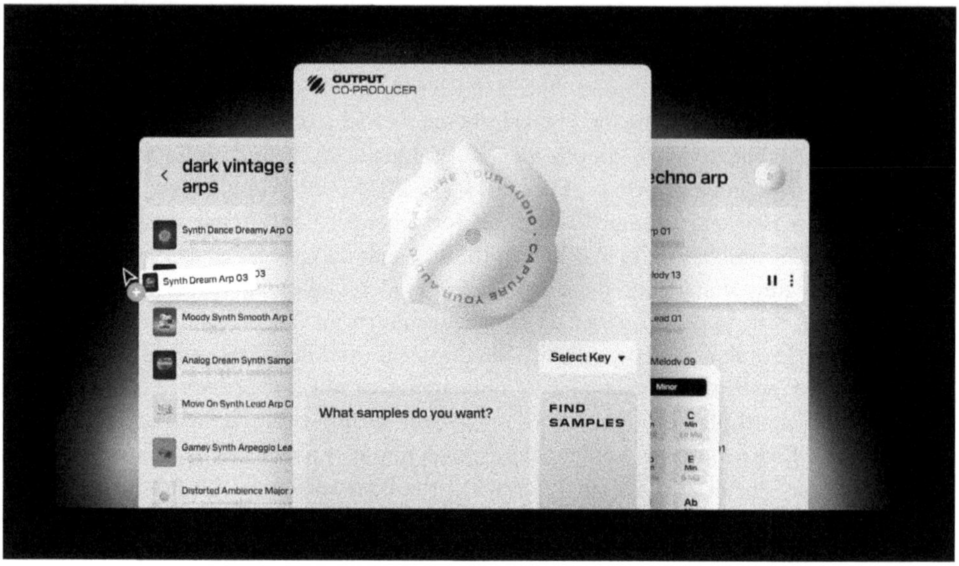

FIGURE I.2 Output's Co-Producer AI plug-in

"The real issue is often how to design or find a sound you're hearing in your head or think would fit with other elements...It just feels like too much tech trying to solve the wrong problem" (lateanonymous4742, 2025).

In the meantime, if there is a production problem in need of solving, we might imagine future text-based AI tools that are more adaptive to a user's input. As Output's Co-Producer plug-in foreshadows, perhaps DAWs will soon have built-in AIs that can respond to a prompt (e.g. "Build me a sample pack based on my field recordings of dumpster sounds"). Or one will be able to select a MIDI chord sequence in the DAW and then use a prompt to specify how to modify its chords (e.g. "jazz harmonies in the style of Bill Evans"), develop their rhythmic profile (e.g. "hocketing like a Balinese gamelan"), and so forth. In a 2024 Hacker News discussion about the AI music generator Udio, an amateur producer named mrandish envisions such a future for interactive and responsive AI tools:

> Don't generate finished songs. Instead, make an interactive tool which collaborates with human music makers in a much more granular way by generating the elements and components of music (called stems) as well as the underlying MIDI data. There's a whole industry selling human-created element libraries consisting of stems, loops, backing tracks, samples and style-based construction kits. These are used in *a lot* of the human-created music we hear. But they aren't interactive, adaptive or collaborative.
>
> *(mrandish, 2024)*

In contrast to text-based AIs such as Suno, Stable Audio, and Diffusion, mrandish imagines an interface that combines two forms of user input: text prompts, and MIDI chords/phrases provided by the user. Together these inputs could generate "a variety of suggestions in the form of separate stems tracks with MIDI which [a producer] can further mix and modify." In sum, just as experience is the producer's perfect plug-in, a useful AI collaborator of the future would be "always available, non-judgmental, infinitely patient" (mrandish, 2024).

Three case studies: Actress, patten, Lee Gamble

As of this writing, some producers have begun incorporating AI in their workflows. A common use case is using AI to generate samples such one-shot drum hits, virtual voices, or chord progressions that become bespoke building blocks for tracks.[5] Let's consider three case studies.

Actress, *Young Paint*

Producer Darren J. Cunningham (aka Actress) began experimenting with AI in 2018, on his EP *Young Paint*. Young Paint is the name Cunningham gave his AI learning program with which he collaborated to make the recording. On Young Paint's Spotify page, Cunningham explains that the program "has been progressively emulating the Grayscale to Silvertone process Darren J. Cunningham started" since his first recording, *Hazyville* in 2008. While the workings of Young Paint aren't specified, Cunningham says that he provided the software choices or "mode selects" of genre (e.g. pop, classical, world), impression (e.g. ballad, joy, groove), and composition style (e.g. tension, movement, fluidity). With different combinations of mode selects as constraints, Cunningham then fed the AI a "detail script"—perhaps some of Cunningham's tracks?—upon which the software generated "its own musical speech" (Bruce-Jones, 2018). Cunningham poetically describes the *Young Paint* EP as "investigating the vacant soul of computer language" (Bruce-Jones, 2018) and frames his AI experiments in terms familiar to producers—as an inevitable man-machine interface. "People become digital. And it's the same with music" he says (Pemberton, 2019).

patten, *Mirage FM*

Mirage FM is a recording by producer and visual artist Damian Roach (aka patten). Roach made the music using sounds from Riffusion, software that generates audio based on user prompts. Over a long sound-design session, Roach tried out many prompts and recorded Riffusion's responses. He then edited and processed the audio into brief tracks. *The Guardian* calls Roach's music "a dizzying headspace" (Beaumont-Thomas, 2023). Nodding to both the DJ practice of mining old records for sample material and AI theoretical concepts, Roach calls his workflow "crate-digging in latent space". Like hunting through old records in search of breaks, there's a hit and miss aspect of collaborating with AI:

> A lot of what was being spat out by Riffusion was not great. You had to wait a long time for something, a tiny little thing that made you say 'yes, okay, there's something there.' A lot of it was looking for things that had a simultaneous sort of familiarity and also a strangeness to them, that I could kind of tease out and push in the direction I wanted to.
>
> *(Mullen, 2023e)*

There are different ways to hear *Mirage FM*. On the one hand, its tracks have an unusual, lo-fi and artifact-heavy mono sound that lacks reference points. Production-wise, the tracks aren't lush or groovy or timbrally attractive. Their sound is often strange, summed up by one reviewer as "familiar and unintelligible" (Ravens, 2023). One might even hear the tracks as musically malfunctioning because they lack traces of a human touch shaping them. On the other hand, the music challenges the listener to understand its peculiar artifacts, its cyborg voice traces and its wonky beats that never settle. As producer Daniel Lopatin (aka Oneohtrix Point Never) observes, perhaps what's most interesting about these early days of AI for electronic music producers is "seeing how AI fails" (Petrusich, 2023). Roach explains that his inspiration for collaborating with AI was to "interrogate the way things are" in music production, and more broadly, to foreground big questions about making art, such as "How do our minds work? What is creativity? How are ideas made?" (Ravens, 2023). In this regard, *Mirage FM* succeeds in turning Riffusion's unmusical riffs into music. If *Mirage FM* offers little to relate to, it's nevertheless thought-provoking.

After listening to *Mirage FM* I went to the Riffusion website and typed in some prompts, just as I had done with Output's Pack Generator. This time it was "ambient afrobeat", "techno gamelan", and "disco marimba." The results were abysmal, although so too were my prompts. (What prompts did Roach use?) Part of the problem is that different prompts produce barely differentiated results, and no matter the prompt, Riffusion's output, like that of many text-to-audio generators, generally lacks the human x-factor—*musicality*. This is a discussion for another time, but what does it mean to be musical, anyway? If we consider, as sound engineer Robina does, that music "is the phenomenal representation of spirit" (Robina, 2021: 65), then Riffusion's output scores low on the musicality scale.

Lee Gamble, *Models*

Producer Lee Gamble made his 2023 recording *Models* around AI-generated voices. Gamble worked with a programmer who built a neural network-based vocal synthesizer that was trained, over a period of months, on many examples of human singing. Once the voice model could produce decent quality output, Gamble wrote additional musical parts around the vocals. His goal was to find a balance between AI's synthetic sound and more conventional musical sounds. "I wanted to really emote these voices" he says.

> I didn't want them to come across as these so-called computerized distant things. I wanted to see what happens if you have that thing where they're not really people, but they might make you feel like you're getting moved by somebody.
>
> *(RA Exchange, 2024)*

A Pitchfork review of *Models* describes its use of AI as raising "questions about expressive authenticity: How much of an artist's soul comes through in music generated by an intelligence beyond the artist's control?" (Broomfield, 2023). This kind of critique of AI's use in music production and whether or not it dehumanizes an artist's sound is reminiscent of the criticisms of early synthesizers. The sound of the RCA Mark II, for example, was once described as having "a grotesquely inhuman quality that comically defies the very meaning of music" (Canby, 1955). Moreover, Gamble's curiosity about AI is similar to Roach's in that he too describes his project as a kind of production challenge that raises questions. "It's about, 'How much mood can I get out of this?'" Gamble explains. "How human can I make the technology? How human does a singing voice feel when it's disembodied? How emotional can a digital simulation feel?" (McKinney, 2023).

Contemporary practices II: vocal deep fakes, vocal cloning and licensing

The most garish illustrations of AI's entrance into 21st century music production were vocal deep fake tracks on which producers used AI voice modeling software to conjure virtual sonic likenesses of famous singers singing songs that the real singers had never sung. Several examples appeared in 2023, including a producer named ghostwriter977's "Heart on My Sleeve", a fake collab between an AI Drake and an AI The Weeknd, an AI Johnny Cash singing "Barbie Girl" by the 90s pop band Aqua, and an AI Kanye West singing "Hey There Delilah" by the rock band The Plain White T's (Coscarelli, 2023). Delighting in jarring stylistic juxtapositions between voice and content, these examples of sonic verisimilitude evoked the "mash up" tracks of the early 2000s (Mullen, 2023f). Artists, however, took umbrage at AI faking their voices. Recalling the critique of the RCA Mark II synthesizer, singer-songwriter Nick Cave memorably called AI songs generated in his likeness "a grotesque mockery of what it is to be human" (Cave, 2023). For the moment, the novelty of vocal deep fakes has worn off.

Meanwhile, some producers have taken control of their sonic likenesses by exploring vocal cloning and making simulacra of their voices available for others to use. In 2019, producer Holly Herndon and digital artist Matt Dryhurst created Spawn, "a singing artificial intelligence baby" trained on the voice of Herndon and others. Herndon used the software as one choral voice among many on her 2019 recording, *Proto* (Friedlander, 2019). In 2021, Spawn was followed by *Holly Plus*, a clone of Herndon's voice that lets others make music with it (Herndon, 2021). In 2023 producer Claire Elise Boucher (aka Grimes) did something similar with her Elf.tech AI platform whose voice software allows users to upload their own vocal tracks for transformation into Grimes' voice. Artists can commercially release the results of their AI collab "in exchange for half of any master-recording royalties" (Monroe, 2023). And in 2024, producer Imogen Heap, known for her creatively pitch-shifted vocal harmonies going back to her 2005 hit, "Hide and Seek", developed ai.imogen, an AI voice model intended to transform any instrumental or vocal input "into a passable simulacrum of Imogen Heap" (McEvoy, 2024). While it's not clear there's much demand at the moment for such production ventriloquism, it's likely that in the coming years more musicians will license AI versions of their artistic personas. Instead of downloading sample packs, one might use an AI tool to *create vocal harmonies in the style of Imogen Heap*, and so on.

Conclusion: lessons from AI in electronic music production

Given the warp speed pace of its development over the past few years, AI will continue making inroads in music production workflows. Whether it's writing counterpoint in the style of Bach, creating samples from text prompts, or cloning voices, AI's capabilities have spurred musicians to re-think many facets of production. But as Deirdre Loughridge observes, one danger of AIs and symptomatic of their logics is what she evocatively calls a corresponding "falling away of the hand". AI logic, she says, "regards anything performable by machines as no longer a matter of human intelligence" (Loughridge, 2023: 176). Three questions then, come to mind.

First, in light of what AI can already do, what remains of the musician's *necessary* labor and the time it requires? What processes, if any, can or should be outsourced to AI? Take track mastering, for example. In 2023, Apple's DAW Logic Pro incorporated an AI-powered device, Mastering Assistant, that uses machine learning to analyze the spectrum balance of a track and automatically apply EQ and dynamics processing. The device isn't a replacement for a trained mastering engineer, yet it can quickly—in seconds!—produce results that are adequate for many musicians. Summing up the quandary AI potentially puts musicians into, Matteo Wong argues that getting quick results is AI's superpower. "The programs aren't necessarily doing something no human can" he says, "they're doing something no human can in such a short period of time" (Wong, 2024).

Two connected aspects of the music producer's labor are repetition and decision-making. Composing music is a repetitive, iterative exploration-exploitation process of trying out and fitting together sounds and structures in the search for expressiveness. Indeed, creative process lives in the space of our relentless trying: the more we make, the better sense we have of what makes a quality track. But as Joshua Rothman notes, another danger of AI is that, by removing labor from our process it "can relieve us of the burden of trying and trying again...The whole painful cycle of trying, failing, revising, judging, and redoing can be replaced with something simpler: plucking the most suitable result out of a pile" (Rothman, 2025).

And when our labor is diminished so is our power to make aesthetic decisions. As producer Cristobal Tapia de Veer warns, "AI is the next step for deleting uses you have for your brain" (Hanging out with Audiophiles, 2025). While the arrival of AI may mark a turning point in how artists create, the works we recognize as creative are so because of the perceptible skill that went into making them. This is what the novelist Ted Chiang argues when he says that it's the decision-making embedded in craft that distinguishes real art from AI-generated facsimiles:

> Art requires making choices at every scale; the countless small-scale choices made during implementation are just as important to the final product as the few large-scale choices made during the conception. It is a mistake to equate 'large-scale' with 'important' when it comes to the choices made when creating art; the interrelationship between the large scale and the small scale is where the artistry lies.
>
> *(Chiang, 2024)*

A second question about AI in music production concerns the caliber of AI's musical outputs. Much of today's AI-generated music, like patten's *Mirage FM* discussed above, has a noticeably inhuman or *uncanny valley* quality. In other words, by sounding almost human it elicits eerie feelings (Mori, 2012). For a careful listener, today's AI sound contains audible indicators that alert us that something's off. For me, it's an absence of taste, a dead-eyed emotional vibe, an artifact-ed sound quality, a *I can't quite put my finger on it* sense of unmusicality. It just *feels* weird. The AI uncanny valley sound evokes an automated voice narrating a magazine article that occasionally places the emPHAsis in a curiously WRONG place. Surely though, as AI models improve, the AI sound will become more difficult, if not impossible, to notice. Or maybe the AI sound, if we can still speak of one in a few years time, will have become so naturalized that we will no longer hear it as odd.

And a third question is what can music producers learn from AI? One way to answer this is by considering some production lessons embedded in AI tools and the ways we're already using them.

Taste is paramount. So far, AI has entered music production primarily by way of text-to-music generators that are proposed as shortcuts for sound designing and composing. Is this a good thing? Notice that it's CEOs of technology companies, and not music producers, who claim that using AI in music production is net positive because producing is too much work. For example, describing the process of searching for drum samples, Splice CEO Kakul Srivastava says "you just go through and you listen to these sounds…That is a painful process and it shouldn't be that hard" (Patel, N., 2025). Or Output CEO Gregg Lehrman claims that "endless scrolling for sounds [holds] back creativity" and that producers "want sounds that fit their track from the start" (Roche, 2025). Or Suno CEO Mikey Shulman argues that music production as it's currently practiced isn't enjoyable because of the skills it requires. "It takes a lot of practice" he laments. "You need to get really good at an instrument or really good at a piece of production software" (Stebbings, 2025).

Srivastava, Lehrman, and Shulman miss the point: musicians' sense of taste comes *from* practicing skills. And while the proliferating text-to-sample AIs can generate sounds in any style, the tools' outputs, for the moment, tend to be generic, sometimes strange, and lacking taste. Unlike real musicians, an AI model has no taste because it isn't sentient—it has no lived experience, no ability to ground its musical symbols "in real world objects or events" (Patel, D., 2025: 37). It behooves producers then, to double down on their taste to pursue their valuable omnimusicality. For producer David Guetta, this means "more and more [songwriting in the AI era] is going to be about taste, and not only technical abilities" (Hiatt, 2023). To have taste, as producer Guy Brewer (aka Carrier) explains, is to leverage one's experience and thereby become one's own filter:

> I try to think of myself as a filter, especially when I'm using equipment that's semi-generative or there's heavy processing to a point where I'm not really in control of the output. Even in those circumstances, I'm still the one who decides when to hit record, and the one who stitches things together. That all happens through the filter I've developed over the years, which is inherently personal.
>
> *(Reynaldo, 2025)*

Experience is embodied intelligence, a life's worth of knowhow, a music production collaborator with us at all times, attuned to our methods, an intuitive assistant who filters, who helps us make connections between one sound and another, who can say *If you like that sound, you might also find this one interesting.* Our experience knows us well, summoning what needs to be recalled when needed. Experience pushes us to try new things, extends our mind and the tendrils of our thinking, helping to enrich and accelerate our music production journeys. And unlike a musical AI based on LLMs but with zero experience of the world, *our experience is earned by doing.* Accruing experience gifts the producer with finely tuned tacit knowledge, intuition, technique, a sense of craft, lucid thinking, and the ability to apply lessons learned from yesterday to today. Our sense of taste is our style, our take on things. And as producer Chris Clark (aka Clark) says, "having a take on something is the most important thing in music" (Hanging out with Audiophiles, 2023a).

Train yourself on your own datasets. As a thought experiment, imagine yourself as a large language model. Think broadly about how to train on specific influences—even your own previous work. How might you refine your artistic algorithm? Think about your learning model. How would you generate your data, find patterns in the data, and use what you learn to generate ideas (cf. Loughridge, 2023: 166)? How would you inspire yourself?

Think about prompts in the context of music production. Just as the output of AI text-to-audio generators is shaped by our text prompts, we can make prompts to orient ourselves *before* embarking on a project. As a phrase to focus our attention around a process or desired result, a prompt is similar to a template in that it can direct our workflow towards a future output. A prompt can be as vague as

> play this instrument for 30 minutes,
> conjure something oceanic,

or as specific as

> make a 140 bpm beat with dubby echo effects, layered with slow vocal lines and long chords.

Or a prompt can inspire sonic worlds that don't yet exist:

> music for wind instruments and white noise percussion,
> long harmonic sequences that become a resonant space,

and so on.

There are no right or wrong prompts, just prompts that get us thinking. Even uninspiring prompts are useful because they model what *not* to do. (Recall Stable Audio's unimaginative "calm meditation music to play in a spa lobby" prompt.) So devise a prompt and then answer its call. Notice which prompts hem your work in, and which ones lead to surprising journeys. Committing to a prompt can propel a production workflow.

Think about creativity as artifact. AI LLMs can be unreliable, generating inaccurate, fabricated information. When an LLM makes errors, apparent in its "nonsensical answers to factual questions" (Chiang, 2023), it does so because of what is known as a *hallucination* from its training data. Ted Chiang describes AI's hallucinations as the text equivalent of the lossy compression artifact of a JPEG image whose quality degrades each time the file is re-saved (Chiang, 2023). With this in mind, think about your own creativity as *an artifact of processes behaving in surprising ways.* As we saw in Chapter 1, *serendipities* is a broad term to describe such surprising behavior of musical processes. In sum, AI's hallucinations remind us that *our creativity in music production is an artifact of accidental discoveries.*

Notes

1 For example, Johannes Ockeghem's *Missa prolationum*, a five-part mass scored for four voices written in the late 15th century. The piece is structured using a form called *prolation* canon, which means that the same melodies are sung by different voices at different rates of speed. We explore prolation canons more in Chapter 8.
2 The term stochastic comes from the Greek *stochastikos* ("skillful in aiming"). It was first used in 1662 by theologian John Owen.
3 Eno credits Reich in a conversation in Reich's *Conversations* (Reich, 2022).
4 In a 2014 article in *The Atlantic,* William Hochberg wrote about how computer scientists "are working on algorithms enabling computers to make up original fugues *in the style of* Bach, improvise jazz solos *à la* John Coltrane, or mash up the two into a hybrid never heard before" (Hochberg, 2014, italics added).
5 For example, in 2024 the producer Christian Brunn (aka Virtual Riot) released a six-hour stream on his YouTube channel in which he uses the AI text-to-speech generator ElevenLabs to source sounds for a track (Virtual Riot, 2024).

5

SONICS AND INSTRUMENTS

Introduction

On YouTube there's a video from 1996 of the producer Rupert Parkes (aka Photek) demonstrating how he sequences drum and bass breakbeats, using the DAW Cubase to trigger sounds from an E-mu hardware sampler. As Parkes clicks in MIDI blocks on the screen that trigger the breakbeat samples, he credits the E-mu. "The sampler's the main instrument" he says. "I pick sounds for the texture in them" (prestonloyola, 2008). In an interview a few years later, Parkes explained how drum and bass is "a style of programming, a style of using sonics to make music, as opposed to using instruments to make music. This was music purely about soundscapes, space, bass" (Cardew, 2020).

This idea of using sonics as opposed to instruments to make music has long been a key principle of electronic music production. But in the genre's early days, pioneering technologies and their associated sounds received a chilly reception. Listeners and critics were suspicious of synthesized sound. In his 1955 *Harper's* magazine record review of *The Sounds and Music of the RCA Electronic Music Synthesizer*, Edward Canby called the synthesizer "a device designed to eliminate the entire apparatus of musical performance, by people and instruments, that has been with us since the beginning of the art" (Canby, 1955). Canby voiced ideas about electronically synthesized music that have persisted ever since. In Canby's view, the problem with the synthesizer, a "product of pure engineering", is that it lacked both inventive composers for it and a repertoire that could stand on its own:

> The machine, having no composer-of-the-future and no new musical language to operate upon, is forced to use present stock and, absurdly, to go about imitating the very instruments and performers that it is supposed eventually to supersede. A fine contradiction!"
>
> *(Canby, 1955)*

By the time producers like Parkes were sequencing breakbeats and "using sonics to make music" electronic music had found its composers of the future, apparatus of performance

DOI: 10.4324/9781003585145-7

(the computer, the DAW), and a new musical language upon which to operate. Today, the core of production is an obsession with sonics and ways of achieving them. Sonics signify—they tell us about how a music is supposed to feel, what it's intended to mean, and relate one music style to another. Producers' choice of sonics also tell us about their taste, their sense of what's musical. We use our sense of taste when we listen to the tracks of other musicians, asking ourselves: *How did they get that sound?* As producer Jon Hopkins explains, "When I hear something that's been made electronically of course I have a vague sense of how it's made, sometimes. I love it when I don't" (Quiet Details, 2024).

Producers achieve their sonics by many methods, including MIDI programming, sound design, sampling and field recording, synthesis, effects processing, and playing with presets. Our methods don't matter as much as our enthusiasm for sounds we deem musical. Musical sounds can be realistic-sounding, or not. Producer Danny Wolfers (aka Legowelt), for example, is a connoisseur of old, crappy-sounding reverbs:

> I'm kind of more into old reverbs, especially 12-bit reverbs that don't sound too realistic. Modern reverbs that are too realistic are a little bit scary for me because you have a kind of uncanny valley that sounds so realistic it feels uncomfortable. It's nice when the reverb is a bit crappy—it's more relaxing for the brain.
>
> *(Voltage Labs, 2022)*

By contrast, producer Austin Cairns (aka r beny) is drawn to the almost-acoustic sonics of physical modeling synthesizers, such as the Korg Z1:

> I love the uncanny valley acoustic sounds that physical modeling synths produce and the Z1 is just an absolute gem in that regard. For myself, physical modeling is not about replicating the sound of true-to-life acoustic instruments, but rather creating patches that have acoustic characteristics.
>
> *(Quiet Details, 2024)*

This chapter explores ideas about sonics and instruments in electronic music production. It examines sound feels and vibe, instrument agnosticism, virtual versus acoustic instruments, stacking simple tools, ear candy/incidentals/subliminal little sounds, sonic artifacts, the appeal of analog sound, and the alchemy of imperfection.

Sound feels

Every sound has its own feel. This feel is primarily a function of a sound's timbre, or what is metaphorically referred to as tone *color*. In the world of acoustic music, timbre is shaped by an instrument's materials and the techniques for playing it. Timbre communicates to us that it's a flute we're hearing, not a violin, a low piano tone and not a gentle timpani roll.

The timbres of different musical instruments have distinctive feels too. The timbre of a *shakuhachi* bamboo flute is airy, while the timbre of brass crash cymbals is piercing. In electronic music, a sine wave is a mellow and one-dimensional timbre, while a sawtooth wave is more textured and edgy. (The corresponding wavy and jagged shapes of sine and sawtooth timbres are confirmed in oscilloscope representations of them.) Timbre is also shaped by how forcefully an instrument is played. It depends on a musician's touch: at louder dynamics the *shakuhachi* sounds strident, while cymbals played softly sound mysterious. This brings up a question we'll explore later in the chapter: If an acoustic musician's sound feel is transmitted through physical touch, how are elements of technique and touch transmitted in electronic music making, where one's gestures are typically at some remove from the resultant sounds? For now, we can say that producers engage with touch by thinking about how sounds feel at every step of their workflows.

Whether making the heaviest of beats or lightest of ambiances, among producers' many omnimusical skills is attuning to the many subtly different ways a track can feel.[1] We sense the music's feelings as fragments of information that are clues to next production steps. Listening to a track in progress we sense that

it feels dead
it feels busy
it feels dark
it feels ecstatic
it feels ambivalent
it feels like it has no chill
it feels unfamiliar
it feels hopeful
it feels simple
it feels alive.

Some of these feelings suggest fixes: if something feels too busy, just remove material. But responding to the feel of one element and adjusting it can damage the music's overall feel—with some of its parts removed, now the track is less busy but too simple. The producer doesn't know this until tinkering to the point of making the music sound worse and only then realizing how its various sound feels are interlinked. The overall feel of a track, once we're attuned to it, is a compass for getting our bearings within its expressive world. In sum, *the most important attribute of a sound is its feel.* Sound feels matter because making us feel is what music is designed to do.

From feel to vibe

Connected to feel is the concept of vibe. One of the mysteries of music is how it conveys the feelingful energy of the musicians who make it. Short for vibration, vibe poetically describes music's most intangible yet palpable quality—its feeling. Vibe is how music feels over time, how its vibrations sustain a mood, conjure an aura, keep a sensation aloft.

Describing a music's vibe is difficult because it requires assigning a vocabulary to sounds which are immensely varied, nuanced, and therefore elude description. In some ways, music articulates vibes that are uniquely embedded in musical practice, yet outside the reach of talking about music. As ethnomusicologist Charles Seeger theorized in his 1976 essay, "The Musicological Juncture", "gaps found in our speech thinking about music may be suspected of being areas of music thinking" (Seeger, 1976). Maybe it's because of this gap between talking about music and making music that our talk is often reduced to vibe generalities. *That band has a great vibe*, we say. *The track's vibe is dark*, or *The beat has an intense vibe*. The futility of using words to describe vibe's multivalence—*great* vibe, *dark* vibe, *intense* vibe—points to vibe's value. Pay attention to vibe in music and ask yourself, *What is this vibe and what is it saying?* Vibe, like sound itself, is invisible yet with physical effects—like a current, vibe keeps music powered. More than technical know-how or musical cleverness, vibe is music production's most important thing.

Instrument agnosticism

If you're a musician, you're probably devoted to a single instrument you've learned to an extent that it feels expressive. But a music producer's devotion is diffuse. To borrow from Seeger, this devotion is an area of music thinking with more than one object. It's in this way that musicians who compose by producing are instrument agnostics in the sense of being compatible with a variety of instrumental possibilities. And since electronic music production, as Photek noted, is more about the primacy of sonics than how those sonics are generated and what instruments/technologies were used to make them, producers are generalists who explore myriad ways of making music that sounds acoustic, electronic, sampled, synthesized, or a mix of all of these. *Whatever works.* This is why production is such an engaging composing practice: it celebrates a plurality of sounds and a plurality of methods by which to create new musical worlds.

What *is* an instrument in music production? As we explored in Chapter 2, a producer's musical system spans the acoustic and the electronic, the analog and the digital, the real and the virtual. A musical system is replete with sounds, devices, and configurations on the verge of instrumentality, capable of expressiveness, and worthy of devotion. The scope of what constitutes an instrument in production is as wide as our thinking about the matter. A DAW is an instrument, but so is a synthesizer plug-in. So listen to options wherever they suggest themselves. In sum, instrument agnosticism's lesson is *embrace any sound that's a playable sound.*

Production's plurality can be illustrated by the various ways of making a simple pad sound. We can work additively, beginning with a sine wave and build the sound up from there by layering texture, overtones, and ambience effects. Alternately, we can work subtractively, beginning with an already textured sample—what the producer Fred Gibson (aka Fred again..) refers to as starting "from a point of real intricacy"—and then remove from it to "find order" and reveal a shape the way a sculptor chisels down a stone (Doyle, 2021). Filtering out the sound's texture, overtones, and ambience using EQ could leave something sine wave-ish, but with intricacy vestiges still felt underneath. Or we can make a pad from a field recording—a rainstorm, say—and devise a way to tune its pitches into consonance, perhaps using a notch filter. In production, the routes to arriving at a sound destination are many. This is what producer Kevin Parker (aka Tame Impala) means when he cryptically says regarding sound design, "You don't need something to get something, you can get there some other way. You just have to look" (Danz, 2023).

Virtual instruments

Introduced in 1999 by Steinberg, the first VST (virtual studio technology) instrument was *Neon*, a simple polyphonic subtractive synthesizer with two oscillators, three waveforms, an ADSR envelope, a resonant filter, an LFO, and a detune knob (Steinberg, n.d.). While *Neon* wasn't powerful by today's plug-in standards, its simplicity belied what it heralded: a gradual shift in how producers make sounds with virtual instruments. In the quarter-century since *Neon*'s release, VST instruments have become more capable by magnitudes and are now most producers' go-to sources for making sounds. Virtual instruments integrate numerous types of synthesis, effects processing, and sampling, making them sound design powerhouses. While many of these instruments convincingly emulate the analog music hardware of yesteryear, production's leading edge is arguably with software that celebrates its digital uniqueness and makes possible sonic worlds impossible by any other means. Let's consider three such instruments: Xfer Records' Serum, U-he's Zebra, and Slate & Ash's Landforms.

Created in 2014 by producer and programmer Steve Duda, Serum was a wavetable synthesizer considered the successor to Native Instruments' Massive, which was released in 2006. Updated in 2025, Serum 2 offers a vast selection of wavetable types and filters, four envelopes, an effects rack page, and ten LFOs mappable to any parameter on the instrument (Figure 5.1). Serum's widely imitated click-to-drag-this-to-modulate-that functionality allows a producer to see modulation options and routings on a single page. Serum is inspiring to use: its easy-to-navigate graphical user interface (GUI)—which has been called "one of the most elegant interfaces a complex synth was ever given" (Nagle, 2015)—has made the instrument a favorite of producers, who use it to craft aggressive basses, crisp plucks, glassy leads, and ethereal pads.[2]

Released in 2003 by U-he, Zebra is a modular synthesis environment with additive, subtractive, FM, physical modeling, and wavetable synthesis options as well as numerous

FIGURE 5.1 Xfer Records' Serum 2 synthesizer

types of modulation and effects. As with Serum, Zebra's patching and routing of sound generators and modulators is displayed on a single page. The instrument's audio path has a pleasing visual logic, flowing from top to bottom, without the chaos of a hardware modular synthesizer's tangled wires. By conveying a sense that one could build almost any instrument inside its environment, Zebra inspires its users to pursue the possibilities of synthesis.

Landforms (Figure 5.2) is a sampled library of acoustic strings and woodwinds by Slate + Ash that contains hundreds of recorded textures, from long sustains to quick pulsations, wavering undulations to gossamer harmonics. Landforms' sampled sounds are performances replete with micro-variations; playing just a single note generates an ever-changing, continuous texture. The instrument's sound design options include tuning, envelope, microphone perspectives, expression, and vintage signal processing effects. Last but not least, Landforms' minimalist GUI, which is truly abstract even by virtual instrument standards, makes the instrument look like a Rothko painting.

Serum, Zebra, and Landforms invite the producer to explore them by interacting with their possibilities and serendipities. The more we work with a virtual instrument the more we understand that *what makes the instrument musical is how we interact with it.* As with playing an acoustic instrument, this interaction is influenced by the software's

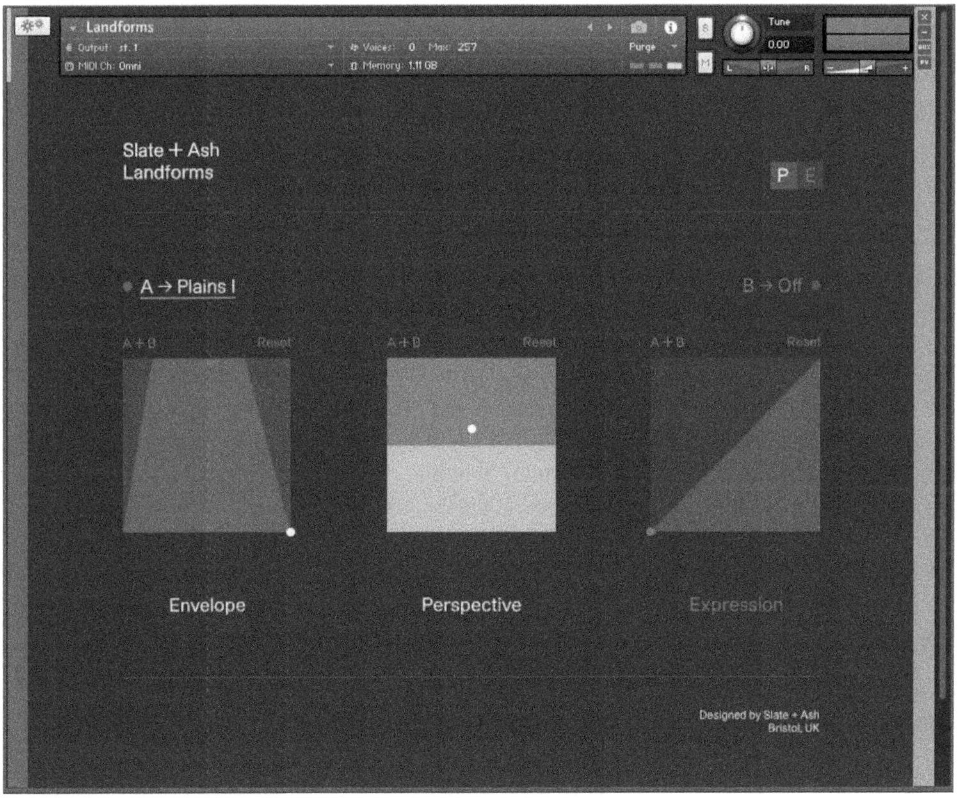

FIGURE 5.2 Slate + Ash's Landforms sample instrument

GUI, the feel of its sounds, the fluidities and resistances it offers, and the decision-making it encourages. In sum, let's consider four qualities in virtual instruments:

It does one thing really well. There's value in single-function virtual instruments. Some software achieves this by mimicking acoustic instruments. For example, Modartt's Pianoteq uses physical modeling synthesis to uncannily evoke a piano's sound production and includes controls for adjusting minutiae of tuning and sympathetic resonance. Other software emulates classic hardware, like Roland's TR-808 drum machine. Or as with Zebra, it re-creates the functionality of modular synthesizers but without the physical limitations of these machines.

It has a GUI that makes visual sense. Ideally an instrument doesn't hide its functionalities. This encourages users to figure it out, more or less, by exploring its pathways of action. Serum exemplifies this accessibility. Producer Bill Day (aka Mr. Bill) recalls that it took users of the original Serum a few years to figure out what he aptly calls the instrument's *meta game*—that is, taking its pathways of action to a next level of inventiveness. "In 2018, 2019, everyone just figured out the meta game for Serum and it became like this thing that you can make anything in. I feel like that will happen for Serum Two as well" he notes (The Mr. Bill Podcast, 2025).

Its parameters are automatable. What virtual instruments lack in sonic realism and user-friendliness they make up for with their parameters' MIDI-mappability. As discussed in Chapter 2, editing and controlling via careful automation is second only to live performance as a way to generate moving details within a part to make it sound alive at many levels at once.

It's unfathomable. Just as we sense ineffable qualities or spirit in the playing of another musician, perhaps we notice something like soul in software, even though its coded logics are not our embodied ones. This is what Patricia Wolf means when she says:

> I like to think of instruments as divining devices that can reflect our thoughts and feelings back to us and perhaps say what words cannot or show us what we haven't quite yet come to understand cognitively.
>
> *(Fischer, 2022a)*

Sometimes there's a disconnect between action and result when we use virtual instruments in that, since we never *physically* connect with them, we don't achieve the input-output connection the way we do when we play an acoustic instrument. But an upside to this disconnect is that it makes production interesting—sometimes even magically so. As we miss sounds intended, unintended ones appear along the way that we can pursue. This is part of what Day refers to as a virtual instrument's *meta game*. Dave Gamble, the esteemed programmer behind DMG Audio plug-ins who also contributed code to Serum 2, observes that an instrument's design offers hidden options for ways of routing its components. "I think there will be great music that comes of it" he says. "There's all kinds of trickery that you can pull off in there that even we haven't figured out yet" (The Mr. Bill Podcast, 2025). The soul in virtual instruments lies in their unfathomability, their untapped potentials a perpetual reminder of how much we could be pursuing.

Sound is a moving target: playing virtual instruments

The oldest basis of making music is having a relationship with an instrument. This relationship comprises touch, technique, and reciprocity. An acoustic instrument is something physical you can touch, get vibrating, and make sounds with that responds to the quality of your touch and sounds in accordance with your technique. In doing so, the instrument gives something back: it reciprocates by helping you make music, and also by resisting your efforts. Fingers on the fretboard or keyboard have to work with, and around, this resistance. "Physical instruments" computer scientist and musician Jaron Lanier reminds us, "channel the unrepeatable process of interaction" (Lanier, 2023). Without an instrument to play and interact with, musicians (notwithstanding singers, whose voices *are* their instruments) can't do much.

Skilled musicians have a relationship with their acoustic instruments deep enough that an instrument feels like an extension of one's body. The drum and drumstick extend the drummer's hand, the violin and its bow allow the violinist's fingers to sing, the trumpeter's horn and its mouthpiece amplifies breath into tone. Having a relationship with an instrument means learning to play at the interface point at which you and the sound-making machine meet and interact. Musicians earn this connection through experience. It takes years of practice to get to know how one's instrument works and how to work with its workings. Through lessons and practice one learns tone, technique, repertoire, and how to play this repertoire in a stylistically appropriate way. It takes even longer to feel like you can *express* yourself—to feel you're one with what is in fact a mechanical contraption that's physically separate from you. Over time, the musician develops a tacit knowledge of an instrument's expressive potentials, a sense of what can be done with it.[3]

Now let's turn from playing acoustic instruments to playing virtual ones. If acoustic instruments are extensions of our bodies and require a tacit knowledge of their expressive potentials that takes time to develop, is this also the case with virtual instruments? And what characterizes the experience of playing one? To return to the question we asked in our sound feels discussion earlier in this chapter: How does a musician transmit technique and touch in electronic music making?

Obviously, the musician-instrument equation changes in the context of electronic music production with regards to both virtual instruments and hardware ones. Instruments such as synthesizers and samplers are physical, but generate their sounds by electronic and digital circuitry. Circuits can produce enchanting sonics, but the main difference between playing a synthesizer and playing an acoustic instrument is that, with a synthesizer or a sampler, one's sound isn't consistent. Its basis can be a waveform or a sample or synthesis; change a few parameters and the sound morphs into another, different sound. Or a sound can be a hybrid of several sources interacting. With a non-acoustic musical instrument, one's sound is a moving target that potentially never stops moving. Virtual instruments multiply one's options for generating timbres. When one combines virtual or hardware electronic instruments with effects processing, there's no end to devising bespoke, digital hybrid instruments without equivalents in the acoustic world. Because of the optionality of music production, a producer can feel the lack of a consistently meaningful relationship with a sound world. So how does one establish this relationship?

One way a producer creates consistency is by *playing*. If we can drum or play a part on say, a keyboard controller, we have *an approach to the sounds we're working with* that's grounded in our history with the controller's corresponding acoustic instrument. Not every producer is a keyboard player, but keyboard controllers are accessible (and perhaps more beloved than grid matrix controllers) because the piano has a 400-year history and is an instrument with which many musicians have some familiarity. Playing parts connects producers to the varied sound sources in their musical systems, transmitting some of their touch and tacit knowledge into a track. As producer Kelly Lee Owens explains, recording ourselves playing an instrument is a way to leverage our relationship with it and thereby bypass the DAW's endless optionality:

> Think about it. When you're programming stuff into the computer, you are constantly assessing what it sounds like and whether this note needs to be changed. It's a constant interruption because that's how your brain works. But *if you create a piece of music on an instrument and immediately record that*, you are taking away your brain's ability to doubt and question what it's doing. Once it's in the computer and you've moved on to the next bit of the song, that's it...no more worrying if you've got the right hat sound."
>
> *(MusicRadar, 2022, italics added)*

Let's consider four additional points regarding a producer's relationship with virtual instruments. First, *when we play a virtual instrument, our touch doesn't determine the sound we make.* When we play a MIDI controller that controls a virtual instrument, we're not *making* a sound, but rather *triggering* one. This means that some of our musicianly know-how about sound production is blunted from the get go. When we play a controller, our sense of touch doesn't get to directly touch the music. It's in this way that playing a virtual instrument misses the magic of acoustic musicianship.[4]

Second, *when we play a virtual instrument and record the results, anything we play is provisional* in the sense that we can change it after the fact. We fix wrong notes, re-shape dynamics, and re-order pitches to make new chords long after we played them. MIDI is malleable, and finessing it is a way to retrospectively inject touch into the music. Most significantly, we can change a virtual instrument's sound after recording a part. This is how a lead melody reappears as a bass line, or piano chords as a string orchestration.[5] As we explore in Chapter 8, generating multiple instrumental lines from a single MIDI part is one of the producer's most powerful practices because it allows for trying out options.

Third, unlike a musician's longstanding relationship with an acoustic instrument, a producer doesn't necessarily stay with a virtual one for long, instead switching among instruments constantly, practicing what we referred to above as *instrument agnosticism.* We try new instruments and then circle back to old ones, comparing the functionalities of each, waiting for an *Ah-ha!* moment when it becomes clear that this is the one we're going to commit to, at least for now. And since virtual instruments are updated, the old becomes new again, over and over. Alternately, we might limit ourselves to one instrument (as a minimalist constraint) if we find a path forward with it. I once made a sample instrument out of a long harmonium tone, and then recorded a few tracks with it.[6] Maybe I'll return to this harmonium sound someday and use it as a building block for something else.

A fourth point about playing virtual instruments is that *our relationship with them consists of slowly building a memory of their uses.* This memory may be more visual than kinesthetic—like a remembering of which controls and parameters are where and how we've used them to make interesting sounds. The producer Christian Valentin Brunn (aka Virtual Riot) explains how, prior to the era of YouTube production tutorials, he learned to play the VST synthesizer Massive by turning every one of its virtual knobs to figure it out:

> Just turn every knob, remember what it does and start to build in your mind a sort of map of, *Okay, this does this, this does that, now I'm hearing this sound which has this sort of characteristic, and if I put this knob here it would get* [a sound] *like that.* So that's what I did for years.
>
> *(Virtual Riot, 2023a)*

As discussed in Chapter 2, a producer's memory for a virtual instrument's uses is helped along by our *sound fishing* practice. Whenever you work with an instrument, try to save at least *one new sound*, with the goal of accumulating a library of favorites you can return to. This library of sounds is our tacit knowledge externalized in the software. It's also a gift to our future selves: when we re-open an instrument, we'll be surprised to see how much we saved in its library. (What sounded cool then, tends to still sound cool.) Over time, producers remember generalities of what they've done with this instrument or that effect. (*There's a bunch of wobbly pads here…I made some interesting grainy delays with this…*) Saved sounds and effects can also be connected to make instrument-effect hybrids. These chains add to a producer's repertoire of instrument-effect combinations.

The relationship we have with virtual instruments is our embodied knowledge remembered: As we work with the software, we build a sense of the musical things it can do. This is a different kind of musicianship than that required to drum, or play piano, but not completely different. When we play an acoustic instrument it's a physical encounter with a resistant machine, and when we play a virtual instrument it's a non-physical encounter with an opaque system of relations.[7] Both ways of making music require a commitment over time to alchemize oscillation or synthesis into a beautiful, beguiling sound. The power of an acoustic instrument is that it can only be the instrument it is, while the power of a virtual one is that it can be almost anything we want. In sum, the potentials of virtual instruments are a beckoning from the future musics we haven't yet figured out how to make.

While sound for the producer may be a moving target, there are some reliable ways to find sounds we deem expressive. One way is to *design a sound that you enjoy playing.* Sound designing takes time, has no guaranteed results, and is an activity that's different headspace-wise from playing and arranging musical parts. But maintain a sense of urgency—once you've arrived at a sound, try making music with it right away! This is what Harold Budd emphasized in Chapter 4, when he speaks of focusing "entirely on a small thing and try to milk that for all it's worth."

Another way to find expressive sounds is to *refine a sound with effects.* I use effects to color a sound and set a mood conducive to a way of playing. For example, I'll add bits of delay and reverb effects to create a sense of space. Such simple effecting changes how I play—I use fewer notes, letting the reverb and delay add tails to the pitches so they linger and bounce around in time. Refining a sound with effects often makes playing it feel more expressive.

This leads to a third way of finding expressive sounds: *re-use a track's existing elements as the basis of new timbres*. For example, use an existing sound to make an oscillator's wavetable, or import it into a sampler to make an instrument (e.g. a Tibetan finger cymbal becomes a drone). None of these techniques promises useable sounds, but trying them and following wherever they may lead forges a connection with whatever results from the process. In sum, no matter what virtual instruments you choose and which sounds you make by whatever methods, strive to develop a meaningful relationship with what you're hearing. It will happen! Sounds are moving targets, but when you love a sound, it stops moving for a moment and agrees to collaborate, offering you a chance to make something consequential with it.

Stacking simple tools

Electronic music producers strive to make sounds that are emotionally moving in subtle, complex, or enchanting ways. A route to achieving this is by stacking simple plug-ins and running a track's sounds through them. *Stacking* means to connect serially (one after another), and *simple tools* are conventional effects and processors such as reverbs, delays, EQs, bit crushers, compressors, among the numerous devices available in DAW software and beyond.

Stacking simple tools is a non-technical, and therefore—as Brian Eno might say—elegant way to produce intricate sonics. Try the technique on any sound in a track by experimenting with different processing combinations and follow what sounds good. Stacking's magic happens when multiple plug-ins create conditions for sonic serendipities—when the plug-ins combine in unforeseen ways to generate unforeseen, cascading composite interactions. It's the surprises of the musically not-yet-heard that stacking simple tools enables us to pursue.

A way of practicing stacking is to process a drum pattern's component sounds. Start with a generic kit and then process your way forward, optimizing for interestingness (Graham, 2023). With a beat in hand, experiment with treating its individual percussive voices with delays, LFOs, and filters to alter their rhythms and timbres.[8] Maybe the snare gets echo and a moving comb filter, while the kick is glitched and re-pitched here and there. Once you've processed individual drum sounds, you might treat the kit's entire soundset in a similar way, compressing everything and then cutting high frequencies. Experiment to find results. Multiple rounds of sound-shaping may produce something so unlike what you began with as to render *where* you began unimportant. In sum, stacking simple tools allows the producer to start with one form and transform it into something very different. Keep stacking and processing until you're in a place you don't recognize but want to stay in for a while.

Ear candy/incidentals/subliminal little sounds

The sweet metaphor is not accidental: ear candy are brief sounds added in and around a track to sweeten it, moments of interestingness that bring sparkle, razzmatazz, and energetic pop. Ear candy are details finessed, differences that make a difference in how the music feels. In an insightful 2008 video tutorial, producer Christopher Mercer (aka Rusko) used the term *incidentals* to describe these details:

> Once we've got the snares and kicks in there, sounding nice and fresh, I put a few *incidentals* in—extra sound effects to give it that sort of shuffle. I've got a piece of filtered white noise for the start of the bar. It sort of thickens [the texture] up a little more, kind of fills around the top end.
>
> *(Computer Music, 2013)*

Produce your own ear candy to fill out a track, generating the bits from inside the music, or outside. Accentuate artifacts already in a sound by boosting their noise with compression, or highlighting frequency filigree with extreme EQ. Alternately, layer samples to make a patchwork of timbre-textures: add hand claps to join the backbeat a smidge ahead or behind, buzzing transients that hover high above a sub bass, a white noise riser cascading one section into another.[9] Don't doubt the music's confection; we build to delight our ears.

Producer Kieran Hebden (aka Four Tet) talks about a kind of ear candy sounds he calls *subliminal little sounds*. Hebden credits producer William Bevan (aka Burial) as an innovator in this regard. Bevan is known to place "tons of subliminal little sounds and textures all the time" in his track arrangements (Tape Notes, 2024b) to create what critics have aptly called *hauntological* musical interest (Fisher, 2013). In his own work, Hebden uses two subliminal little sounds techniques. The first is doubling one sound's part with a different sound playing the same part. Hebden creates the second part through trial and error by "taking the same MIDI notes and just trying loads and loads of different presets" (Tape Notes, 2024b). For example, on his track "Storm Crystals" (2024), he doubles a bass sound with a bell:

> There's this bell combined with the bass, which just makes the notes stick out a little bit more—you can't really hear the bell too much. It's sort of subliminal, but without it, it's burying the bass more in the mix.
>
> *(Tape Notes, 2024b)*

A second Four Tet'ian technique is to recycle bits of vocal samples and textures already in the track and place them into an arrangement as echoes of the main parts from which they stem. "I use a lot of things like that" Hebden says. "Where there's little quiet echo-y bits of vocals and noises and little things happening all the time" (Tape Notes, 2024b)

Ear candy and subliminal little sounds are music production equivalents of what computer scientist and animator Ed Catmull calls *microdetails*. Catmull, a co-founder of Pixar, brought a scientist's rigor to the problem of how to make enchanting animated art. Catmull's thinking resonates with producers trying to make music alive at many levels at once. Microdetails, he explains, are tiny elements that inform the look and feel of a work on an almost subliminal level, "a hidden engine" that lends crafted objects authenticity and conviction (Catmull, 2014: 198).[10]

Sonic artifacts

Artifacts are details, the hidden gems inside a sound you don't fully hear until you excavate them. Producers are artifact seekers. "It's finding details in things: that's what really interests me" says producer Mark Clifford of Seefeel (Tone Glow, 2021). *All* audio contains sonic artifacts hidden within, and the way to reveal these artifacts is to amplify them, inviting them to be the music's focus, to come center stage and sing their song. Amplify artifacts by boosting their volume or EQing them to highlight what's already there, to bring forth latent sounds within the sounds. Other second-order artifacts may appear as by-products of processing already applied. For example, when a distortion or a reverb leaves traces of its micro-grit texture or subliminal hum, our dramatic gain-staging can bring this background to the fore.

Producer Geir Jenssen (aka Biosphere) often uses samples of acoustic instruments—especially orchestral recordings—to anchor his tracks. As we discussed in Chapter 2, it's the artifacts in these sounds that become interesting when amplified and processed. The late producer Brian McBride, of the duo Stars of the Lid, also composed with acoustic sound sources. McBride reveals that the source of the "strange noises" artifacts in the music was acoustic guitar, strings, and horns recorded and then re-sampled:

> It's all about layering instruments and making new instruments out of the combination of different instruments' decay and timbre. But the important part for me is that everything is real. I don't use any sound banks or keyboards. I do have a MIDI keyboard but it's just for playing my own sounds.
>
> *(Krasowski, 2008)*

Producer Brian Leeds (aka Huerco S.) metaphorically likens the producer's artifact-seeking process to that of the archeologist who locates fossils deep below ground, cleans them, and then shares the discovery: "Listening to samples or synths and layering them, brushing off the dust, finding the artifact, then presenting it to the public...I'm just going to keep tinkering away, conducting experiments, and discovering artifacts deep deep below" (Arnott, 2013).

Like Hebden recycling his vocal samples and textures, Leeds derives his artifacts from what is already in the track. He foregrounds them because they sound interesting:

> It's not like I'm adding an entire track of white noise or static. *It's always things that are present, and I just choose to accentuate those things instead of hiding it*—I just let those things shine through. Like low-bitrate samples. You hear the compression in the audio, these artifacts. I mean, I shouldn't be using these bad things...I think they sound interesting.
>
> *(Miller, 2014, italics added)*

The appeal of analog sound

From around the 1950s to the early 1990s, when most music was recorded either in studios or concert halls, its continuous sound waves were captured by microphones, sent into a mixing console, and recorded on a magnetic tape recorder. While the musical sources could be acoustic (e.g. an orchestra) or electronic (e.g. a synthesizer), the recording path was always analog in that it was a continuous electronic signal flowing from microphone to mixer to tape. The signal's voltage varied continuously, in step with the pressure of the sound waves produced by the music being recorded. Over a roughly 40-year span of time, a so-called golden age of audio production unfolded, during which now-classic albums (e.g. by the Beatles, Miles Davis, Steely Dan, Michael Jackson) captured not just first-rate performances but also an analog circuit path whose sound is still admired today.

Analog can describe several facets of contemporary production. The first is the above-mentioned *signal path*, whereby a sound source is captured by a microphone and then passes through electronic hardware before being recorded. Another facet of analog is the *electronic circuit-based synthesizer*, such as those introduced in the 1960s and 70s that produce sounds through voltage-controlled waveform generators, electronic oscillators, and filters. A third and perhaps most contested facet of analog is a *sound quality* that some musicians claim to be able to discern. Analog's sound is often described as warm, in contrast to the cold sound of digital. There's a material basis for this aesthetic distinction. A digitally recorded signal captures and represents sounds through a series of discreet samples; a digital recording is a kind of a musical kineograph—a series of tiny photos of sound that produce a constrained and quantized representation. In contrast, analog captures sounds through a continuous representation. The alleged warmness of the analog sound is believed to be a by-product of analog recording equipment that preserves the integrity of sound's inherent continuous variability, as well as the subtle distortions and saturations introduced as sounds flow through this gear (or virtual emulations thereof).[11]

With this in mind, consider the analog purist, who comes in many forms. Some are recordists or audiophiles who insist on using as much vintage analog hardware—e.g. preamps and compressors—as possible.[12] Some purists are record collectors who swear by the tactile sound of analog's medium of choice, vinyl. And some purists are fully digital music producers who, despite working with computers, go to great lengths to recreate analog's sonic *wabi-sabi* aura, either by recording analog electronic instruments or by using plug-ins that evoke those instruments' wonky oscillators and emulate analog hardware signal processors and effects.[13]

In recent years, a turn towards so-called "lo-fi" music can be heard as a kind of nostalgia for some, but not all, aspects of the analog sound. A lot of this music highlights analog's imperfections. Examples of this include foregrounding the low-fidelity sonic artifacts of analog recording technologies, like the hiss and wobble of cassette tapes or the crackles and pops of old LPs. "These imperfections are so important" notes Adam Harper, "not just because they are of interest sonically but also because they are understood as indicators of authenticity, that is, a lack of mediation" (Harper, 2025). The lo-fi aesthetic comes together in "lofi beats" instrumental music, produced in the DAW yet evoking the all-analog sound of a vague yesteryear. Consider a two-minute track called "controlla" by the producer Lassi Kotamaki (aka Idealism). Here is a brittle hip hop beat (which Kotamaki describes as "steady, softened beats"), a little bass, and tape-wobbled piano chords. In other examples, such as Iván Rosa's "Ciudad de Papel", a track's entire mix is run through effects that muffle and distort to create the sensation that all the sounds are old. Related to the lo-fi phenomenon, the rise of "synthwave" music is another form of analog nostalgia. Synthwave uses conspicuously-vintage synthesizer and drum machine sounds, often from the early 1980s. Listen, for example, to Michael Glover (aka Miami Nights 1984)'s "Ocean Drive", a track that evokes the sepia neon aesthetic of a Miami Vice or Flashdance soundtrack.

The point is that analog nostalgia in music production always looks backwards, evincing a mania for the retro (cf. Reynolds, 2011) or at least a wistfulness for a time when recorded music allegedly sounded better—meaning "warmer" and "less than perfect." By strategically using vintage analog sounds, lo-fi and synthwave producers show their interest in selectively signifying an era of recorded music's history. While producers today record digitally, the analog aesthetic is here to stay.

The alchemy of imperfection

If the analog aesthetic is fundamentally appealing because it brings the vibe of imperfection into our work, many production tools nevertheless stubbornly steer us *towards* perfection. We play samples with immaculate verisimilitude to acoustic instruments, auto-tune errant vocal pitches by quarter tones, quantize sloppy drum hits to the nearest sixteenth note or downbeat, loop phrases to exactly four measures, and limit the life out of mixes so they never clip. In short, we're always tempted to jettison imperfection. In 2015, Brian Eno explained the temptation to correct/perfect/quantize/retune music in the DAW and the impact of such digital manipulations on the vitality of a musical performance:

> It's a problem everybody is having at the moment. Do I resist the temptation to perfect this thing? What do I lose by perfecting it? It's difficult. Because now it is possible to mend anything, correct anything. The rhythm's a bit out on that bar? OK, we'll just stretch it a little bit. We can quantize everything now, we can quantize audio so the beat is absolutely perfect...So you're always asking yourself, have we lost something of the tension of the performance, of the feeling of humanity and vulnerability and organic truth or whatever, by making these corrections? It does make you question the role of new technology in the studio.
>
> *(Spice, 2015)*

The problem with such perfection-oriented production moves is that they're *unmusical.* They suck the life out of our expressive gestures, they remove tension and urgency from our performances, and they remove touch. Producers don't want perfection, they want imperfection. But how is this achieved?

A way to imperfection is to stay loose, free, and nimble for as much of the production process as possible. Ideally you want to leave things undefined—not quite knowing what you're doing or where you're headed—for as long as it is practical to do so. This will keep your options open later on. Compose music that has life to it by preserving the imperfections that are inherent in your improvising and recording processes. This means leaving the errant finger drumming un-quantized, and if you must use loops, use ones that surprise. Think of imperfections as evidence of your struggle, the quirks of your unique experience, the *Oh what's that?*, and *I wonder if...?* moments that make your music what it is.

Imperfections are everywhere in production's processes. Recognize them when they arise, embrace them, and fold them back into the work to make it more natural, more like nature. As Christopher Alexander said in Chapter 4, "in nature you've got continuous very-small-feedback-loop adaptation going on, which is why things get to be harmonious." Folding imperfections back into our tracks makes our music more harmonious. Taking this idea even further, think of imperfections as what music is about. Imperfections are significant because they're ruptures in everyday way of doing things and opportunities for thinking anew about what we're doing. Leonard Cohen frames a similar idea in his song "Anthem" when he sings "Everything has a crack in it; that's how the light gets in" (Cohen, 1992). Imperfections are the cracks in our musical experience that make us unique. So pursue the cracks!

Of course, one can also add imperfections at any point to amplify sounds' inherent texture, patina, and grain. There are many plug-ins for degrading sound—from amp distortions to effects that emulate lossy MP3s, spotty cell phone connections, and corroded cassette tapes.[14] Used subtly, these effects bring vibe to a sound in the form of artifacts and nonlinearities, ambiguities and uncertainties. In sum, use sound-degrading tools to make a sound less predictable, dialing in just enough effect to disrupt a sound's message—to add cracks to it. Sometimes, extreme settings can be the most compelling. Sonic disruption makes the obvious thing sound less obvious, moving us from *I know how this goes* to *What is this?*

We conclude this chapter by returning to producer Nils Frahm, who compares sonic imperfections such as tape noise to the artifacts and distortions in old film. Production tools allow us, Frahm says, to shape recordings to be either clean and perfect or dirty and imperfect. He always choses the latter:

> Maybe its a nostalgic or romantic approach because I find tape noise beautiful and it reminds me of my childhood or old movies and things from the past we don't face anymore in our sterile, digital, crisp and clean world. But we can use artifacts and distortion in a more artistic way as something which is on the other side of the middle. We can now choose if we want to have a crystal clear super objective recording or if we want a more distorted subjective recording of things. Now that we can record sounds that are clean, perfect and absolutely in sync, we should think about imbuing the failures and imperfections in the same way people have rediscovered Polaroid or film.
>
> *(Nair, 2013)*

In sum, each track is an opportunity for the producer to seek out imperfections. From staying loose and nimble to degrading sounds, imperfection's alchemy lies in forging something beautiful from the unexpected. Imperfection recognizes that music's damaged and uncertain bits can be its best parts.

Notes

1 In the second stanza of his 1915 poem "Peter Quince at the Clavier", Wallace Stevens makes a simple observation about the nature of music with an acuity that exceeds the findings of music theorists: "Music is feeling, then, not sound" (Stevens, 2011: 90).

2 Christian Valentin Brunn (aka Virtual Riot)'s Serum-focused tutorials on his *Virtual Riot* YouTube channel are a good place to start. See for example, "Studio Time S2 EP3 - Track From Scratch" (Virtual Riot, 2023).

3 To take one example from drumming: drummers and percussionists have a unique understanding of the phenomenon of bounce. The way their sticks bounce depends on the instrument the sticks are striking, as well as the sticks themselves. (Mallets with soft heads on them bounce differently from drumsticks.) The varied surfaces of snare drums, bass drums, timpani, cymbals, gongs, and marimbas each have distinct bounce demands. Some of these surfaces are tensioned (e.g. timpani heads), some aren't (e.g. marimba bars), and each surface in vibration reacts differently to one's sticks and mallets. Drummers then, not only keep time and accent it; they also continuously adjust the bounce of their drum strokes to the ever-changing vibrating state of their instruments. This connects to Chapter 2's discussion of Brian Eno riding on the dynamics of his musical system. When you watch a jazz drummer's right hand keeping time on a ride cymbal, know that the drummer is smoothly riding the vibrations of the cymbal through bounce.

4 However, some recent electronic keyboards claim to be hyper-responsive to our sense of touch. For example, Expressive E's *Osmose* synthesizer, released in 2024, announces "a new standard for musical expressivity" with the instrument's three-dimensional control over each key: "Press, bend, shake, strum & more to interact with keyboards & electronic sounds in a new organic way" (Expressive E., 2025).
5 See composer Paul Thomson's tutorial of how he transforms a piano sketch into an orchestral arrangement (Thomson, 2024).
6 For example, my track "Harmonium in June" (2024).
7 In his classic book *Technique for the Virtuoso Timpanist*, the influential percussionist Fred Hinger observed that all instrumentalists face the common problem of *resistance*. "Resistance is provided partially by the instrument and partially by the performer himself" he writes (Hartenberger, 2016: 17).
8 Ableton Live's Drum Racks encourages this sound-by-sound processing. To process a drum sound with an effect, drag the effect onto the sound's corresponding pad in the Drum Rack (Ableton, 2024: 397).
9 At the time of this writing, Cableguys' *Snapback* plug-in effect does exactly this: automatically layering samples on top of drum hit transients, to bring "magic to grooves and life to digital hits" (Cableguys, 2025).
10 A more historical take on this is Paul Graham's musically-attuned explanation of what makes Leonardo da Vinci's 1475 painting *Ginevra de Benci* so powerful to behold: "All those unseen details combine to produce something that's just stunning, like a thousand barely audible voices all singing in tune" (Graham, 2003).
11 Producer Cristobal Tapia de Veer observes that analog sound is appealing because "pretty much anybody likes fat music, music that feels like you're close to it, and that happens with that saturation and compression. When there's none of that, it feels like daggers coming at you, like it's somehow attacking your ears" (D'Silva, 2023).
12 An anecdotal example: an audiophile friend of mine bought a vintage vacuum tube-based McIntosh amplifier for his vinyl-only home music listening set-up. When I asked him how his all-analog system sounds, he said it was, uncannily, like having a jazz band in the living room.
13 Leonard Koren's description of the Japanese concept of *wabi-sabi* captures the beauties of analog sound as heard or imagined: "Things wabi-sabi are expressions of time frozen. They are made of materials that are visibly vulnerable to the effects of weathering and human treatment. They record the sun, wind, rain, heat, and cold in a language of discoloration, rust, tarnish, stain, warping, shrinking, shriveling, and cracking. Their nicks, chips, bruises, scars, dents, peeling, and other forms of attrition are a testament to histories of use and misuse" (Koren, 2008: 62).
14 For example, AberrantDSP's *Sketch Cassette* plug-in has a switch to select among Regular, Chrome, and Metal bias cassette tapes, as well as different amounts of wear.

6

LEARNING PRODUCING AND BUILDING TACIT KNOWLEDGE

Introduction: from bi-musicality to omnimusicality

In 1959, the ethnomusicologist Mantle Hood published "The Challenge of 'Bi-Musicality'", a short essay about the skills musicians need to be fluent in more than one musical tradition. Hood was a scholar of Javanese gamelan music, who taught at UCLA at a time of increasing interest in non-Western musics of non-European origin—what would eventually be called "world musics". UCLA and several other universities offered students performance classes in gamelan, West African drumming, and the classical musics of India and the Middle East. Hood's essay, an early attempt to explain the specialized training required to achieve competence in an unfamiliar tradition, introduced the term *bi-musicality* to describe the "specific problems which confront the student who is learning a foreign music" (Hood, 1959: 55). Practically speaking, what would it take to become a proficient player of say, the Turkish *ney*, or Indian *tabla*? Hood explained various skills required by music students undertaking non-Western music studies, including a sharpened aural perception (e.g. the ability to discern microtones), learning by imitation and repetition within an oral tradition, an advanced sense of rhythm (e.g. hearing polyrhythms, playing intricate hand drum strokes), and fluency in improvising. Hood suggested that developing such skills, while difficult, is the way to become bi-musical and thereby make an "alien musical expression familiar". "The training of ears, eyes, hands, and voice and fluency gained in these skills" he wrote, "assure a real comprehension of theoretical studies" (Hood, 1959: 55).

Hood's essay was ahead of its time and relevant beyond its world music context because it was an earnest attempt to understand the elements of musicality in terms of a musical tradition's requisite skills. Over half a century later, the challenges Hood outlined for students seeking bi-musicality through non-Western musical performance have analogues in those faced by students of electronic music production. One could even make the case that electronic music is the 21st century's world music in terms of its practitioners' diverse competencies, musical tools, and the genre's broad range of expression. Electronic music production is vast and to learn about it one needs to understand and practice the various fluencies its workflows require.

DOI: 10.4324/9781003585145-8

The equivalent of Hood's *bi-musicality* in music production is what I call *omnimusicality*. Consider production as a unique experience, which asks the practitioner to work with sound by many means, in many modes. "It's multitasking" says producer Tom Jenkinson (aka Squarepusher) explaining how he works, "switching from one mentality and flipping back" (Assar, 2012). Producers are omnimusical in that they sound design, program MIDI, sample and sculpt audio, make beats, shape arrangements, craft mixes, and build tensegrity through automated parameter interaction. They do all this in a quest to create dynamic musics that feel alive at many levels at once. And because producing requires so much of its practitioners, it's a way to develop a panoply of competencies. When we compose a track, we build a feedback loop that propels us forward along a tumbling coil of learning how to learn.

This chapter reflects on strategies and approaches for learning music production as one practices it. It examines finding one's sound, forward engineering, judgment and tacit knowledge, practice as repetition without repetition, repeated listenings, elements of artistry, and error's opportunities.

Where do I find my sound?

On Reddit, a novice producer was looking for help, seeking advice about where to find their sound. "I do not know" they write disarmingly, "where to get the sounds that I hear in songs or how to create them" (Where Do I Find My Sound, 2022.).[1] The musician's question is simple, but its answer is complex. By "sound" the musician means the recognizable style of production, signature techniques, and feel/vibe by which a track identifies its maker. Replies from the Reddit community offered actionable advice. "Find sounds you love and tweak them", "Really try and master your synth and DAW", "Build up a folder of sounds you either love to re-use, or unique sounds you've made" they suggested.

What makes the *Where do I find my sound?* question complex though, is that a music producer's sound has many sources, some of them specific, others vague. If the producer were an acoustic instrumentalist, the source of their sound would primarily be in their touch, meaning their approach to playing an instrument, a topic explored in Chapter 5. But the touch of producers is diffused through the many parts of their musical systems which span musical software and hardware, as well as acoustic instruments. A producer's sound is a function of many choices made over a music-making workflow and in fact over a musical life. As discussed in Chapter 2, producers play a musical system of their own devising as a kind of meta-instrument, a studio as a compositional tool (cf. Eno in Cox and Warner 2004: 127-130; Moorefield, 2010), and adventurous artists use their DAW not as mere multitrack recorder but as a multifold, ever-mutable creative environment. For these reasons, the sound of a producer, unlike that of an acoustic instrumentalist, has more than one source. A producer's practice is *omnimusical* in that their musicianship resides in many places at once—in improvising, composing, sound designing, arranging, orchestrating, recording, and engineering. The producer makes sounds by as many means necessary.

Returning to the novice producer's question, finding one's sound happens as a by-product of becoming an experienced musician. This takes time and intentional work over many years. "I can't impress on you enough" says mix engineer Dave Pensado, "that it is the person's taste and determination, combined with experience, that determines the sound. NOT GEAR" (Pensado, 2021). Pensado's determination component is key, because it's controllable. Building on this idea, we find our sound by intentionally doing specific production things. Let's consider a few of them.

Learn to play your DAW as an instrument for generating sounds you can't imagine until you hear them. The DAW isn't a fixed sound world, it's open-ended, capable of generating any sound you can imagine (cf. Théberge, 1997), even if you can't imagine the sound until you actually hear it. Finding your sound requires innovating your own ways of making sounds. This comes from trial and error, exploring possibilities, and figuring out how to achieve what you find compelling.

Devise ways to imperfectly replicate your favorite sounds. Listening is the starting point for developing our own sound. Trying to replicate what we hear other musicians create inevitably leads us to workflow insights and fascinating sonic byways.

Make tracks out of sounds you accidentally happen upon. Embracing accidentally-discovered sounds by making tracks with them is a way to practice flowing with serendipity and develop a sense for how unanticipated moments can shape one's music.

Build a tacit knowledge muscle memory of key commands. Learning keyboard shortcuts changes our mental map of software environments, so make time to practice them to navigate the DAW with minimal effort. As Kevin Kelly notes, time spent on improving how you do your thing "is the most powerful way to amplify and advance your thing" (Kelly, 2023: 196).

Think of musical results you'd like have, then explore processes for realizing them. This connects to our discussion of musical prompts in the Interlude. Create production goals. For example, ask yourself *How would I make a grainy ambiance? Or a beat that gets increasingly longer?*

Learn your sound by making music using a minimum of elements. Building on Chapter 3's discussion of minimalism and constraints, musical style is as much a function of what isn't present as what is. Therefore, use the fewest number of parts to achieve a sound. As producer Barclay Crenshaw (aka Claude VonStroke) notes, "when you only use seven sounds and they're all unbelievable, and you spend a lot of time on each one of those sounds, it's minimal but it's really good" (The Mr. Bill Podcast, 2023).

Always ask, Why is this sound right, but that one isn't? Being discerning requires maintaining a high threshold of quality. Style is not a function of technique as much as having a stance or a sense of what's aesthetically pleasing and a commitment to realizing it. A connected idea is: reject with gusto sounds that don't work. Always be discerning.

Develop an aesthetic that is yours alone by trusting your taste and following it. We find our sound by making the music we would want to listen to.

Forward engineering

If reverse engineering is analyzing a finished work to figure out how it was built, forward engineering is analyzing a track in progress to figure out what it needs to be developed. This isn't knowing in advance exactly how a track will end up (that would be fortune telling), but rather an openness to being both partially lost while working and attentive as to what might happen next. We don't know where the music is going until it gets going, and we don't envision a sound until it comes into view. For producer Hania Rani, this means having "a huge trust in the sound itself and what it has to offer…collaborating with the matter that is appearing in front of me" (Headphone Commute, 2023b). The kind of forward engineering that producers can implement into their practice isn't magically predicting a track's future, but rather trusting in sound—*collaborating with the matter that is appearing*—and making sensible production decisions based on this to push the track along a trajectory whose endpoint is unknown. This is fine because, as Matt Ridley writes about innovation, it's often the case that "techniques and processes are developed that work, but understanding of them comes later" (Ridley, 2021: 282).

A sensible production decision has one or more of the following qualities:

It helps the music adhere to broad conventions of genre and style, whether those conventions are dance-oriented, atmosphere-oriented, experimentally-oriented, and so on. Note "the sonic markers of the subgenre you're working in" advises one book on techno production (Trimbacher et al, 2024). For example, deciding on and committing to a tempo, a set of rhythms, and an initial sound set locates the music somewhere rather than nowhere.

It incorporates traditional instrumental roles in the music, such as a beat, a bass line, a chord structure, a melody, and so on. Even when there aren't recognizable instruments in a track, synthetic sounds nevertheless often play roles modeled on those of their acoustic instrument counterparts. The principle here is that even music that's otherwise alien-sounding can nevertheless evoke the *we're-listening-to-one-another* interactions of a band of human musicians.

It helps the music cohere. Shared timbres and effects glue the music's parts in subtle ways, bringing unity to what could otherwise be a disparate collection of sounds. For example, multiple parts can come from the same sample, or a single effect on a Return channel can treat several sounds with the same processing.

It provides the music with a temporary scaffolding. When you don't yet know what a part should be doing, repeat it so that its form becomes a scaffolding for other parts. You can return to the repeating part later to vary it, leave it as is, or delete it. Create scaffoldings within a track by repeating parts within a section and then repeating entire sections. This creates ever larger fractal forms.

It recycles, at any point in the process, what's already present in the music. The most potent example of this, discussed elsewhere in this book, is *resampling* or processing an existing track and recording the results to make a new part. But recycling options are everywhere—a moment of audio can be turned into a sampler instrument, or a reverb tail amplified into alternate ambiance. Like the chef who prepares a vegetable in three different ways, producers extract multiple textural options from one set of materials.

It subjects the music to various sonic treatments. Recycling via resampling is a form of what Brian Eno calls sonic *treatments*, which is processing existing sounds to generate new ones, a topic explored further in Chapter 7. Some producers use resampling as a treatment for a track's entire mix by putting a chain of effects on the project's main output. This technique, whose results can be ear-openingly strange, destabilizes the identity of a track-in-progress. In one fell effect swoop we've blurred all of the sounds in a halo of reverb, crushed them into bits, or multiplied them into rhythm swarms. The results of such treatments epitomize forward engineering by giving us a glimpse of what our music could become (if we get over our fear of putting an effect chain on the track's main output).

To make some of the sounds for his recording *Music for Psychedelic Therapy*, producer Jon Hopkins resampled parts already in tracks. Resampling is a process of *field recording within the DAW*, whereby one re-records one part to generate another.[2] The workflow might look like this:

process a track with effects,
resample it onto another track,
then process the resampled track again to create variations on the variations of the original.

"It's possible" says Hopkins as he recalled the genesis of some of his album's atmospheric tonescapes, "that some of these sounds are versions of the original sounds, they've just been taken really far away from it" (Hanging out with Audiophiles, 2023). He might begin, for example, with a recording of an acoustic piano, re-sample the track, and then slow down and re-pitch the copy to make a drone that's unique yet still genetically related. Hopkins explains at length how his resampling treatment triggers an exciting sound design journey that unveils new sonic palettes:

> Something I like to do is resampling whole sections. So that what you're listening to—say you really have got lost in that loop, you like it—you can save it as something and resample and then bring that into a new track and start fucking with *that*. Clumsy analogy: if you're oil painting on a canvas, and your palette evolves and evolves and evolves and you've got your base color for the sky that's just got just tiny bits of everything in it. It's like you consolidate that area of your palette with one color and then you use that as the basis for this whole other thing. But that thing in itself has got like thousands of elements in it...
>
> You might put that whole world you made into an effect that completely renders it unrecognizable, or granulates it or reverbs it in an intensely dramatic way. And then you might print that and drop it by an octave and then suddenly you found this strange well of harmonics that were up there that you couldn't hear because they were too high for your ear to identify. And then suddenly a new melody might appear in your mind because you heard those harmonics at that pitch. And then you might bring some elements of the original loop back. I like following these threads and not knowing whether they'll be good or work or not. Often they are interesting.
>
> *(Hanging out with Audiophiles, 2023)*

It takes chances and learns from its many small errors. At every moment of composing, the producer is slightly lost, taking chances, waging tiny bets, making mistakes, following tiny idea threads, and finding one thing while looking for something else. Your hands alight on a chord you didn't plan for and it sounds good; an effect you made for this sound works even better on another one, so you pursue it; you try out an instrument you haven't touched in a while and find what you need; or the section at the end of the track you're looping becomes your favorite, so you revise the music around the loop. The producer incorporates ways of learning on a whim and embraces the serendipitous discoveries that arise.

In sum, sensible production decisions create outcomes that inspire us to reconsider the potentials in our track-in-progress. Forward engineering is about the elegance of recognizing, working with, and imagining the prospects inherent in what you have. Creative producers don't spend more energy than they need to. They trust that practice generates its own momentum.

Judgment and tacit knowledge

Back to Reddit: another novice producer has just got set up with a DAW and begun making music. But they're noticing the gigantic gulf between the sound quality of their music and that of professionally-produced tracks. The novice wants advice on how to learn to make music in the DAW and how to develop a reliable sense of what sounds good. Fellow producers make recommendations: read the DAW's manual, make a new track every day, learn one virtual instrument so thoroughly that it's easy to make *any* sound with it. This advice is good. I would also recommend studying interviews with musicians for technique/workflow insights (as this book does).[3] But the source of the novice's problem is deeper than this.

Underneath the quandary novice producers notice are two fundamental skills: judgment and tacit knowledge. Judgment is the ability to discern and make considered decisions. When we work on music, it's our judgment—not our gear or our musicianship—that's most important. We're making judgments every moment we play, listen, or shape sounds. Even everyday producing moves, like adjusting the volume of a part, involves judgment. As we listen to a part repeatedly, we make changes whilst trying to discern tiny gradations of difference in those changes. We use judgment to adjust sounds' smallest meaningful bits of information—the *differences that make a difference* (Bateson, 2000: 462)—until the music sounds right.

Our judgment is informed by our tacit knowledge—the intuitive sense we've gained through our experiences, which is hard to verbalize or write down. Tacit knowing was famously defined by Michael Polanyi as "we can know more than we can tell" (Polanyi, 2009: 4). This is the opposite of formal or *explicit knowledge,* which can easily be explained and codified into rules. Like a recipe, for example. The roots of our tacit knowledge are as deep and wide as our experience. Tacit knowing is the foundation upon which the producer intuits answers to open-ended questions. *How do I know when this part is right?* (When it sounds right.) *But how do I know what sounds right?* (Trust your sense of what sounds right).

Because our experiences are different, each of us has a tacit knowledge whose sources are unique and sometimes opaque. But opaqueness is okay—following Polanyi, you don't have to know *how* you know to know *that* you know. An experienced producer makes quick judgments—a sound feels dull, a part lacks interest, a mix needs air—and makes adjustments accordingly. Judgment and tacit knowledge are connected in that you access your tacit knowledge about a craft through your judgment that guides you along the workflows the craft requires.

The quandary the novice faces is that they lack what connects judgment and tacit knowledge—*experience.* Experience requires time spent paying attention to sound, a topic explored in Chapter 3. To use a gardening metaphor, experience is the soil in which both judgment and tacit knowledge grow. You can't rush or control experience (or a garden), but you can choose where to aim your attention. Let's consider then, four methods for attention-aiming:

Make something new each day. Doing and making are the foundation of experience. The producer James Robowtham (aka DLR) explains how "it's these little percentages of learning...or achievements that build up to your great understanding of music and sound" (MusicRadar Tech, 2019). As you build you learn about what you like and what you need to do to get the sounds you want. "Art is itself a research practice" notes the philosopher Alva Noë. It's "a way of investigating the world and ourselves" (Noë, 2015). It's by doing and making at the leading edge of our abilities that we develop skill.

Make your learnings explicit. Build each day, but also make time to reflect on what you're doing. Make notes on which techniques led you to pleasing results. This turns your tacit knowledge into conscious knowledge.

Listen widely and narrowly. If you're not making music, listening to music is the next best way to learn about it. Producer Kieran Hebden (aka Four Tet) observes how "the more you take in of understanding the great music that's out there and the things that came before, it's so powerful. People need to listen to way more music" (Tape Notes, 2024b). Listening widely means listening to musics outside your zone of favorites and familiars. Listening narrowly means listening repeatedly and obsessively to musics to figure out the mechanisms by which they're alive at many levels at once. Together, listening widely and narrowly build a tacit knowledge map of the musically possible.

Work on large musical structures rather than small musical details. Try to finish entire tracks rather than laboring over individual sounds and never finishing anything. This chunking principle underlies foreign language learning pedagogy whereby one practices speaking phrases rather than individual words isolated from their contexts. Like language, music make its meaning in longer chunks. Instead of focusing on musical details, commit to musical gestures. Don't make a loop, make a whole track.

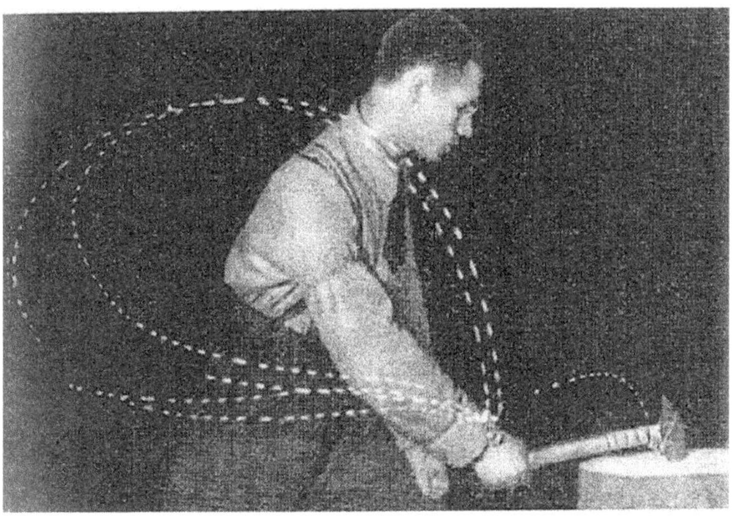

FIGURE 6.1 Cyclogram of a blacksmith's ever-changing movements

Practice as repetition without repetition

In the 1920s, Nikolai Bernstein, a neurophysiologist and pioneer in the field of motor control, studied the movements of blacksmiths at Moscow's Central Institute of Labor. Bernstein used cyclographic photography techniques to track the arm motions used to cut metal with a chisel and hammer (Figure 6.1). His research on blacksmiths showed the importance of *variability* in the subjects' repeating arm movements. Bernstein noticed that the most accurate blacksmiths showed the most variation in their motor patterns, finding ever-changing ways to consistently arrive at the same hitting spot.

Bernstein's research caught the attention of the wider world with the publication of his 1967 book, *The Co-ordination and Regulation of Movements*, a collection of writings on motor control. This book is the source of an idea about the best way to practice: *practice is a particular type of repetition without repetition*. For Bernstein, practicing is a way of solving problems not by doing the same thing over and over, but by subtly varying one's strategies from repetition to repetition—much like the accurate blacksmith who varies his arm/hammer movements each time around. In other words, to practice is to practice variability. In *The Co-ordination* Bernstein applies lessons from his research on motor habits to theorize more generally on what it means to practice:

> The process of practice towards the achievement of new motor habits essentially consists in the gradual success of a search for optimal motor solutions to the appropriate problems. Because of this, practice, when properly undertaken, does not consist in repeating the *means of solution* of a motor problem time after time, but in the *process of solving* this problem again and again by techniques which we changed and perfected from repetition to repetition. It is already apparent here that, in many cases, 'practice is a particular type of repetition without repetition' and that motor training, if this position is ignored, is merely mechanical repetition by rote, a method which has been discredited in pedagogy for some time.
>
> *(Bernstein, 1967: 134)*

Bernstein's ideas on motor habits have been adopted in movement science and sports training, and they're also relevant to artistic pursuits. If we consider music production as a context in which one must repeatedly solve recurring problems of form and feeling, Bernstein's concept is a reminder that producing isn't a mechanical repetition of techniques that worked on previous projects. Rather, it's a fluid practice of using ever-varied techniques to innovate novel ways of making music. Bernstein's emphasis on the process of *solving the same problem over and over, yet in different ways* offers us a holistic view of artistic practice as adapting to ever-changing aesthetic situations. Think about this: a producer repeatedly confronts music with shifting particulars—this sound, these chords, this tempo, my mood, etc. From this encounter they seek some form of interesting outcome.

For novices and expert producers alike, the lesson from Bernstein is to constantly vary the way we work. Since each track is a unique problem, we can insist that we don't re-use sounds, samples, or effects without varying them. This is a way to *practice* innovating by making small changes so that we repeat without repeating. In sum, as each track comes together differently, the producer folds workflow discoveries back into the work in progress. Such discoveries provide a target analogous to the blacksmith's hitting spot, focusing us on practicing novel solutions to the most important recurring problem: *How do I find new ways to make beautiful music?*

Repeated listenings

In a text file I keep an (erratically) updated journal called "What Worked" in which I write notes about music I'm working on. I write to explain to myself how I made sounds and reflect on compositional process in general. One note reads:

> Yesterday I found myself…enjoying some of the music's depths, twists, and turns instead of wincing and wanting it to sound otherwise.

The note goes on to describe the experience of repeatedly listening to bits and pieces of tracks in progress. It's an under-theorized fact that producing involves an immense amount of repeated listening on different levels of scale. We listen

> to the full mix,
> a section,
> one bar,
> one beat,
> part of one beat,
> a fraction of a part of one beat,
> a zommed-in waveform.

Repeated listening reveals tracks in need of adjusting at various resolutions and in multiple places. Tap the Spacebar to play, listen, then tap it again to stop. *Play…stop. Play… stop.* Repeat. Producer Celia Hollander describes her repeated listenings as immersing herself in her sound world's loops. Her process unfolds slowly:

> When I'm working on the computer, sometimes I'll spend time crafting a palette of timbres and make a bunch of loops and then I'll leave it on for hours, and go in the next room and read and hear it through the wall and I'll just leave it on to get acquainted and deeply familiar with it. And then maybe the next morning I'll add the next part. So it takes a lot of time, it doesn't happen in real-time, and listening is the main part of the process and that can be done really directly, like just sitting and listening, but more often it's oblique or with other activities.
>
> *(McNeil, 2022)*

For producers, *listening is the main part of the process*: our repeated listening is necessary and often fascinating. The sounds draw us in—the repetition of hearing the same bits of music over and over induces a trance state of concentration necessary for understanding the presence of our work. Sometimes repeated listening is driven by a fear that our music will never come to life, that it will never be alive at many levels at once. So we get to know the music's sounds on their various resolutions of form—a section, a bar, a beat, a fraction of a beat, a waveform—and make edits to improve what we hear, to dial it in to the best version of itself. Our editing at ever finer *levels of detail* (Brett, 2021: 108) is done with the conviction that, one day, when we hit Play, the music will come alive. Music production's repeated listenings are a labor whose results are proportional to the accumulations of our attentional investments. This is why Hollander says her repeated listening *takes a lot of time* and *doesn't happen in real time.* You get out of your production process what you put into it. In sum, *enjoying some of the music's depths, twists, and turns* may be just one more listen away!

Elements of production artistry

Notice the broad qualities that characterize the work of other producers whose music speaks to you. Maybe it's their inventive sound design, visceral beats, or immersive mixes. Maybe it's their taste, their take on things. Or maybe their entire approach is unusual. As one of many possible examples, let's consider how the work of producer Fred Gibson (aka Fred again..) demonstrates qualities with wider relevance to the study of music production.

Resourceful. Gibson builds his electronic dance music around field recordings captured on his smartphone. Most of this audio consists of voices—singing and speaking—that he turns into melodies and instrumental parts. In his Spotify bio (now deleted), Gibson defined his "Actual Life" production approach as "a trademark sound that features vocals uncovered from the unlikeliest of sources; from obscure YouTube clips to personal FaceTime conversations, random Instagram accounts and iPhone video footage of half-remembered nights out." Given the recording capabilities of smartphones and the ubiquity of social media, it's surprising that there aren't more producers who use what Gibson calls "actual life" samples instead of sourcing sounds from Splice or old records.[4] There's a personal quality to Gibson's sample choices that reveals aesthetics scavenged in unlikely places. Gibson's omnimusical production skill is being an ethnographer of the everyday, who uses ad hoc field recordings to document and recycle the sonic world of himself and his social world. "One thing that's beautiful to me" he says, "is the fact that so much stuff is recorded now, that you can make art out of the actual experience" (Xie, 2022).

Conceptual. Gibson's working with voice samples gathered from everyday life—what he called making *art of out of the actual experience*—is a conceptual approach that insists on adhering to a constraint, a topic of Chapter 3. The constraint is to use everyday sounds captured on one's phone. Gibson notices a pleasing lo-fi quality of phone recordings, specifically how the phone's microphone compresses and over-foregrounds ambiance. "When you record something with your iPhone and stop speaking for a second, it amplifies the room around you" he says. "It has these very crass compression algorithms that I love, so I record everything from drums, vocals, piano, all on my phone" (Ross, 2021).

Gibson's samples don't sound perfect, but what they lack in sonic fidelity they make up for in their social fidelity to the voices and experiences Gibson works with. Brian Eno, who was Gibson's mentor during his teenage years, thinks of the Actual Life production approach as a way of interacting with one's community: "I've actually never heard anything quite like this before. He always seems to be doing it in relation to a community of people around him—the bits of vocal and ambient sounds" (Kamer, 2022).

Inventive. Gibson uses field-recorded voices not just as melodies, but also as source material for making drones. These drones in turn guide the songwriting process. Gibson uses them as *harmonic landscapes* that frame his track's melodies:

> Long drones—different ways of framing what the person is saying. Generally with samples I'll make like five or six drones that it can live in, and the harmonic landscape of those drones will affect the emotion of the song—that's what they're born out of.
>
> *(Lowe, 2022)*

Casual. The most significant quality of Gibson's work is its apparent informality and spontaneity. The way his tracks incorporate snippets of singing and conversations feels intimate and casual, and the music's craft is such that songs don't feel labored. Neither Gibson's dance-oriented nor his ambient-oriented music are built of unchanging loops. Eno, who we've learned is fond of asynchronous loops, appreciates Gibson's non-linear, context-rich constructions:

> What I noticed with Fred is that he would start something, and he wouldn't turn it into a loop that's going to run through the whole track...He doesn't clean everything off, so in every piece of recording, there's a sort of context that comes with it as well. The sound has a history built into it.
>
> *(Dunworth, 2022)*

Finally, the casualness of Gibson's music is mirrored by his minimalist, mobile studio: a laptop, the DAW Logic Pro, and a pair of headphones. The set-up allows him to produce anywhere, and often in transit.[5] Outside the isolation of the home studio, Gibson finds inspiration in the energy of the world. Producing on the move, he says, is about "finding places where you get a conveyor belt of humanity to subconsciously affect you" (Cragg, 2022).

In sum, Gibson's production artistry blends a sense of *spontaneity* as to where he finds ideas for tracks (anywhere) and where he works (nowhere in particular) with a *rigorous* workflow that adheres to the Actual Life constraint. Gibson's is an ecological approach to production that looks to the immediate environment (an Instagram feed, a subway ride) for ideas and, using close at hand tools (a phone as field recorder, a laptop), makes music to capture the feel of a time and place.

Error's opportunities

I'm clicking around an audio file I've loaded into a granular sampler, adjusting the Grain Location to re-position the sample playback at different points along the waveform. Nothing sounds good, so I keep moving the cursor around, now adjusting the Grain Size knob as well. But there's still no sound that inspires me to hear beyond the sample, to *hear through the sound towards a feeling evoked*. There's no magic.

I'm clicking around the waveform because I've committed to use the sampler to make a series of pieces. My plan's vague: *see what you can find in the audio files, then build on that*. So far though, I'm not finding much to build on. Is this idea a dead end, an error?

In his writings about his concept of *antifragility*, Nassim Taleb notes that our errors have "beneficial properties" in that they reveal information (Taleb, 2012). There's an asymmetrical relationship between our small mistakes (i.e. those with little downside) and their potentially big payoffs. For Taleb, making errors tells us what *isn't* working, and therefore helps us understand what we *don't* know. It's through making errors that we learn. In music production, making errors is a way to sharpen our sense for which sounds and techniques work for us, and which ones don't.

Despite there being no magic, I stubbornly continue tinkering with the audio to see where it leads. I open the software's Modulation page and adjust the sample's ADSR (Attack Decay Sustain Release) Envelope. I shorten the Sustain, which causes the sample to skip and stutter. Then I make the Release super long, which turns the stutter into a drone. *What's that sound?*

The drone has a life of its own—its long sustain keeps it going and going, even through I've taken my finger off the keyboard. The drone is full of little pulsations and its sustain is such that I can play a chord in the lower register and, while the chord lingers, add in a few more notes higher up. I record low- and high-register drone parts in separate takes. Recording the parts separately adds unpredictability into the process: since the second (higher) part isn't listening to the first (lower) one, the length of each can be different. The two-part texture isn't something I could play live, and its form is emergent. What felt like an error a few minutes ago may be an opportunity.

One of the most common experiences in music production is encountering our own errors.[6] Among the many types of production errors are those of process, judgment, performance, and patience. Process errors include doing things that we know from experience rarely lead us to interesting results. Errors of judgment include not listening closely enough to what it is about a sound that is or isn't right. Errors of performance happen when we don't take enough risks while playing. In my case, all of these error types connect to a lack of patience.

But since encountering our own errors tells us what isn't working, we can use this information to slow ourselves down and maybe alter our practice. As Nicolai Bernstein would say, we can find new ways of practicing repetition without repetition. My clicking around an audio file in search of something good illustrates this. Errors can be dead ends, but they also present opportunities. Sometimes having the patience to try this and that eventually generates a sound with promise.

Notes

1 For another similar thread, see Prestigious_Gold_742 (2023).
2 In a 2012 interview, Hopkins explains how he even resamples effects channels: "I tend to bounce effects channels into another track and treat them as other instruments, just so you can get their sound very specifically EQ'd, in quite a crazy way if necessary" (Greeves, 2012).
3 These insights are available on the Database page at brettworks.com (Brett, 2025).
4 Along these lines, producer Jon Hopkins explains that in the making of his 2024 recording *Ritual*, he and his collaborators recorded the soundscapes of their studios as they were working in them, in effect capturing, to use Gibson's phrase, the spaces' *actual life* to incorporate back into their music: "I and all the English collaborators recorded ourselves in this room with some surround mics around, and just kind of moved and shuffled around, trying to essentially capture some life, and the energy of the people that made this thing" (Churchman, 2025). Similarly, consider XLN Audio's "Life Field Recorder", a mobile app for capturing 15-second audio

samples on one's phone. The audio cloud syncs to the Life DAW plug-in, which beat-slices the samples into rhythmic grooves (XLN Audio, 2025).

5 An article in the Guardian notes how "much of the Actual Life albums were made while on the move, be it via long train journeys, or on meandering tube excursions" (Cragg, 2022).

6 "Cataloguing of negative results" Taleb advises in the "Heuristics to Maximize Antifragility" section of his essay, "Understanding is a Poor Substitute for Convexity (Antifragility)" (Taleb, 2012a: 3).

7

AMBIENCES AND ATMOSPHERES

Introduction

Electronic music producers design sounds and compose music for these sounds, but they also craft the virtual spaces their music lives in. Ambiences and atmospheres is the principle of bringing dimensionality to a track to achieve the sense that the music is not only alive at many levels at once, but realistically inhabiting some kind of environment. There are many techniques and tools for creating ambience and atmosphere, from sound design choices to reverb treatments, from decisions about how to organize a track's parts to spatially balancing a mix. This chapter explores ambience and dimensionality in music production through the topics of ambient sound's history, using field recordings to escape flat virtual space, drone-making techniques, and the affect of effect chains.

A history of ambience

The idea of conceiving of music and environmental sound as ambience has a hundred-year history. We trace the idea back to 1917, when the composer Erik Satie coined the term *musique d'ameublement* (furniture music) to describe his repetitive, humdrum music, which furnishes an acoustic space by functioning as background sound. Satie's *Forged Iron Tapestry* and *Phonic Floor Tiles*, brief compositions that were just a few measures long, were intended to be repeated indefinitely, mask other non-musical sounds "without imposing itself", fill awkward silences at dinner parties, and create an emotionally neutral mood (Cox and Warner, 2004: 63). With *furniture music* Satie had conceived a looping music as environmental enhancer.

This sonic enhancement idea was formalized in the 1930s by the American Muzak Company, which began using recently developed radio technology to deliver music to workplaces with the goal of altering their soundscapes. Muzak innovated a technique for reshaping popular songs into bland background music that wasn't overtly distracting and functioned on a subliminal level to increase worker productivity.[1] In the 1940s, French radio engineer and composer Pierre Schaeffer conceived his *musique concrète*

DOI: 10.4324/9781003585145-9

(concrete music) radio art pieces, using tape recordings of environmental sounds. His best-known piece, *Étude aux chemins de fer* (*Railroad Study*), features numerous train sounds edited together to create a sequence of railroad-themed soundscapes. While Satie and the Muzak Company shared the goal of creating background music that didn't distract their listeners, Schaeffer's work was oriented around the process of listening itself, proposing what the composer called *reduced listening* "that ignores the source and origins of sound" (Demers, 2010: 173). Recalling Sarah Davachi's advice in Chapter 3 that we listen to the sound of a single oscillator, Schaeffer's ideal listener would suspend disbelief to focus on the "sonorous object" (*l'object sonore*) as "pure phenomenological experience" (Demers, 2010: 175).

Soon after Schaeffer's early pieces for radio, the phenomenology of listening was the whole point of *4'33*, a 1951 composition of silence devised by John Cage, whose use of chance procedures we discussed in the Interlude. The emptiness of *4'33* proposed a radically inclusive idea of music, offering listeners an opportunity to attend closely to any and all incidental environmental sounds that might occur during a performance of the piece. From Cage's Zen Buddhism-influenced perspective, all sounds are equally musical, and *4'33* expressed a philosophy of taking seriously their significance.

Cage's dissolution of the distinctions between musical foreground and non-musical background, Muzak's bland environmental music, and Satie's emotionally neutral *furniture music* informed the *ambient* music concept of Brian Eno, whose music production philosophy we've learned about throughout this book. But Eno's telling of ambient's origin story doesn't cite the innovations of Muzak, Cage, and Satie as inspirations. Instead, he says the idea occurred to him in the early 1970s, while he was recovering in a hospital. He was listening to a recording of harp music playing at low volume and noticed its sound mixing with the rain pattering outside his bedside window. It was the blurring of these two sound sources, Eno says, that gave him the idea of composing a gentle tonal music that would not demand one's full attention but instead be "a place, a feeling, an all-around tint to [one's] sonic environment" (Cox and Warner, 2004: 96). In the liner notes for *Music For Airports*, Eno carefully defined ambient music as sound that "must be able to accommodate many levels of listening without enforcing one in particular; it must as ignorable as it is interesting" (Eno, 1978).

Interest in recording and analytically listening to the ambient sounds of nature coalesced in the late 1960s, when field recordists began documenting the acoustic sound worlds or *soundscapes* of urban and rural environments. Led by the Canadian composer and writer R. Murray Schafer (who coined the term *soundscape*), this research conceived of the soundscape and its audition as a grand musical composition, recalling Cage's notion that music can be any and all sounds. Schafer's World Soundscape Project catalogued diverse sonic environments on stereo recordings to encourage the study and preservation of endangered sounds.[2] Several publications emerged out of this research, including Schafer's *The Tuning of the World* (Schafer, 1977), a theoretical and practical guide to the everyday sounds of rural and urban acoustic environments, and works by fellow project members Barry Truax and Hildegard Westerkamp.[3]

The work of Schafer, Truax, and Westerkamp led to the development of the interdisciplinary field of acoustic ecology, "a composite of scholarly, artistic, and political approaches for studying sound's relationship to the environment" (Demers 2010: 121). It also inspired the practice of soundscape composition, a form of electronic music containing field recordings of specific locations, whose aim, notes Truax, is to "re-design the soundscape, and reawaken people's perceptual appreciation of its importance" (Truax, 1996: 53).[4] The legacies of the World Soundscape Project's recordings and the field of acoustic ecology have been carried forward in the practices of sound artists and academic researchers documenting the acoustic life of various geographic locales, and in the works of electronic music producers.[5]

In the 1970s, the field recordist Irving S. Teibel released a series of soundscape recordings on his Syntonic Research Inc. record label, adapting the documentary and analytical approach of Schafer's World Soundscape Project for therapeutic ends. Teibel's *Environments* series comprised eleven extended field recordings (two side-long tracks per LP) of ocean waves, forest streams, rainstorms, winter blizzards, animal and insect sounds, and so on. In pursuit of a hyper-real and immersive listening experience, Teibel enhanced his recordings like an electronic music producer would, using EQ filtering, overdubs, and synthesized tones. The subtitles of the *Environments* records were intended to sound scientific (and a little psychedelic) by alluding to the records' intended restorative function. For instance, the cover of *Environments 9*, a recording of Pacific Ocean and Caribbean lagoon sounds (Figure 7.1), reads: "turn your hi-fi into a psychoacoustic device" (Teibel, 1976). Other releases in the collection promised "a new easy method of relieving tension" and "totally new concepts in stereo sound" (Teibel, 1971).

FIGURE 7.1 Syntonic Research's *Environments* recordings

Teibel's *Environments* recordings inspired similar collections in the following decades, including the *Solitudes, Echoes Of Nature, Magic Moods*, and *Sounds Of Our Planet* series.[6] Numerous New Age music releases in these series combined nature soundscapes with electronic musical textures to make a sound intended to be "relaxing, calming, and meditative" (Hall, 1994: 23). New Age recordings situated electronic sounds in nature and gave nature's soundscapes an electronic soundtrack. A notable early example of this juxtaposition was Wendy Carlos's *Sonic Seasonings* (1972), which mixes synthesizer parts with nature field recordings. In the liner notes, producer Rachel Elkind recalls Satie's *musique d'ameublement* and even anticipates Eno's ambient concept. She describes *Sonic Seasonings* as "designed to be a part of the décor, so to speak—a sonic ambience that enhances the listener's total environment...*Sonic Seasonings* take listeners out of their environment and into the countryside of their fantasy" (Elkind, 1972).

A half century after the documentary research of Schafer's World Soundscape Project and Teibel's *Environments* series, soundscapes are easy to come by online, and the ubiquity of portable digital recording devices (not least the smartphone) has spurred archives of environmental sounds.[7] One prominent website is freesound.org, a collaborative project founded in 2005 and built around Creative Commons-licensed audio. True to its name, freesound is free, and anyone can browse, upload, and download sounds to use in their work. Freesound is vast: among its sounds are thousands of musical instrument samples and nature soundscapes. (The keyword "rain" turns up some 1500 recordings.) Producers use freesound to find audio sources for new sampler instruments.

Field recordings, escaping flat virtual space, mixing electronic and acoustic sounds

Since most music production happens inside the computer, the producer's antidote to this antiseptic situation is to make field recordings to incorporate into a track, infusing the music with an acoustic soundscape. Field recordings have a long history in ethnomusicology as documents of music cultures and place;[8] they were also a cornerstone of *musique concrète*'s early sampling practices. As we learned in Chapter 6, producers such as Fred Gibson (aka Fred again..) unwittingly draw on these traditions of practice when they make recordings of everyday life to use in their tracks. Moreover, producers' field recordings go beyond sampling, taking cues from film's *Foley* sound effect tracks (e.g. mysterious footsteps, wind), named after Hollywood sound effects artist Jack Foley. Field recordings lend a cinematic, dimensional sense of being immersed within a place. It behooves every producer then, to explore their local environment by recording its sounds and playing with them. Field recordings teach us to listen analogically, to find connections between the acoustic and synthetic sound worlds in production. Hear wind as white noise, thunder as sub bass, wind chimes as sine tones, birds as counterpoint, and rattling subway rides as long drone trips.

One of music production's ongoing challenges is how to make recordings that sound as alive as live music. Producer James Holden explains this challenge by contrasting the vivid, spatialized sound of an acoustic jazz band recording with electronic music's flatter, virtual space:

> Electronic music exists in this flat, digital, virtual space behind the speakers. It's not like when you listen to an amazing recording of a jazz band on a great hi-fi, and you

> can hear that goes there, the bass is at the back, the drums are over in that corner... you can hear the room around you. But in electronic music, the only room you hear in it is very often your own. It's all just pushed to the front of the speakers. It's nice to flip that idea around.
>
> *(Mullen, 2023)*

One technique Holden devised to flip around the idea that electronic music lacks spatiality is to output a track's parts through speakers positioned around his studio and record (resample) the results. This technique, he says, provides "this real spatial sense of what's going on acoustically" in the room (Mullen, 2023). Separating the sounds via speaker playback spatializes a track's parts, creating a sense that they're in a shared acoustic space. On his aptly titled 2023 recording, *Imagine This Is a High Dimensional Space of All Possibilities*, Holden incorporates this spatiality by placing the track's various musical elements at different locations along the mono-stereo spectrum. On the track "Contains Multitudes" for example, the main beat and melodic sequence are mono, but additional layers are added to the mix at increasingly wider positions in the stereo field. At 1:17 a backwards-sounding keyboard enters, and then at 2:34, live *tabla* drumming. Six minutes in, what was mono has grown into stereo and the effect is one of sonic multitudes swirling in a room around the listener, alive at many levels at once. "Contains Multitudes" is extraordinary for the sense of sonic vividness it creates.

The quest for sonic vividness inspired producer Jon Hopkins to explore acoustic re-sampling in a way similar to Holden. For his 2021 recording *Music for Psychedelic Therapy*, Hopkins made spatially-rich field recordings to weave into its tracks. One source was a trip to the Tayos Caves in Ecuador, where Hopkins recorded birdsong, rushing water, and a drone played on a Bluetooth speaker in order to capture cave acoustics. Another field recording source came from collaborations with fellow producer Dan Kijowski (aka 7rays), who played some of Hopkins' audio files through eight speakers suspended in trees in his backyard and recorded the results. Such re-sampling techniques for capturing tone and place also create specific perceptual effects. Hopkins explains that his aesthetic goal was "to move the consciousness of the listener from inside to outside, to feel that space. There's nothing you can do with processing or effects to impart that on a sound" (Currin, 2021).[9]

Producer Ricardo Villalobos' view of ambience is that electronic music will never have the same sense of spatiality as an acoustic music recording. For this reason, acoustic recordings are the standard by which Villalobos calibrates his work.[10] He makes his point by explaining the problem with using reverbs to approximate the experience of being in a resonant acoustic space:

> What I try to do with electronic music is try to sound as good and free and as spacey as an acoustic recording. This is the big reference for me, but it's nearly impossible. Because the electronic frequencies are more defined and the rooms, you have to create rooms with spaces and reverbs. All the other effects you can use as something abstract, but the reverbs...
>
> Humans are experts in reverbs, it is the one thing we are really experts in; we know if we're in a kitchen, an elevator, a big hall. Microphone recording lets in this space where the sound was recorded, but with electronic music, it is an effect that is

> light years behind. You can't betray the human ear, which is why dry recordings can be timeless and reverb recordings are always fixed to the time of a technology. It's impossible to bring into electronic music the space existing in acoustic recording, but we can make field recordings and mix electronic and acoustic music. This is what I try to do constantly.
>
> *(Ware, 2007)*[11]

Many producers share Villalobos' enthusiasm for mixing electronic and acoustic music sources in their tracks with the goal of sounding *as good and free and as spacey* as an acoustic recording. What compels them is discovering ways of achieving atmosphere, mood, and a sense that the music is alive at many levels at once. Producers believe that field recordings imbue a track with a vividness, aura, and intentionality impossible to achieve with sound processing. And field recording captures the timbrally nuanced sounds of acoustic instruments with fidelity, an approach memorably summed up by Hopkins' production dictum, *less laptop, more reality* (Turner, 2013). Now let's turn to examples of ways producers work with field recordings.

For producer Richard James (aka Aphex Twin) a track's "magical ingredient" is the sound of an instrument recorded in its acoustic space. James, who has worked with MIDI-controlled pipe organs and Disklavier-controlled piano, explains that "when something moves through the air, it's automatically going to sound more interesting." This principle even applies to samples: when a sample is played through a speaker and re-recorded, he says, "it's been in the real world and put back into the computer. It's a nice thing to do" (Sherburne, 2014).

Like James, producers Sam Shepherd (aka Floating Points) and Lucinda Chua are fans of processing the electronic sounds in their tracks with amplifier and reverb effects to situate them in the real world. Shepherd's approach is intended to compensate for the fact that

> electronic instruments never existed in the acoustic domain until they leave the speakers for the first time. So I'm really keen on the idea of treating electronic instruments as if they were acoustic instruments. Putting them through guitar amps for instance, giving them a room to exist in. And I think psychologically that does a lot for listeners. Suddenly this synth-line has got a weird reverb sound to it, you get slap-back or whatever in it, it doesn't sound right, but it exists. It manifests itself.
>
> *(König, 2019)*

Chua's processing approach is to automate the amount of amping and reverb effect to create either subtle or extreme dimensional shifts in which sounds move from one virtual space to another:

> I do a lot of reamping and reworking of sounds. Then in the edit, it's very much about automating between those different choices. I do the same with reverbs: rather than just sticking to one I like to be able to dip into different sound worlds and different physical spaces. Sometimes that's a seamless transition from a really intimate sound into a really expansive sound, and sometimes it's a very hard cut.
>
> *(Mullen, 2023d)*

Geir Jenssen (aka Biosphere), who often builds tracks around samples of acoustic instruments, uses field recordings as subtle atmospheric context. "I often feel that even having a really low field recording under the track made in the studio changes the atmosphere completely, you know?" (Howells, 2016). Similarly, Bjarki R. Sigurdarson (aka Bjarki) records acoustic sounds in nature that he later uses as percussion samples:

> Making field recordings is just trying to capture a certain mood. A lot of the sound that you hear—drums, percussion—I would like to capture a certain vibe from nature by just going out with a drumstick and trying to find a kick drum in the stones, or trying to find the kick drum in the sand or lava rocks.
>
> *(Sigurdarson, 2018)*

Producer Klara Lewis thinks of her field-recorded found sounds as both instruments and atmospheres. "I've always thought of found sounds as my instruments" she says. "I use them to make my rhythms and melodies, and build my tracks around them as if they were other instruments." As atmospheres, Lewis's field recordings bring "the outside world and your relationship to the world into the music…It adds a personal element to the work that I think I'd have a really hard time working without" (Hall, 2017).

Producer Felicia Atkinson incorporates into her work recordings of herself whispering words in French and English. On the track "Thinking Iceberg" (2024), Atkinson's ASMR-like whispering lends a sense of mystery and drama, demonstrating the affective "power of hushing" as a sound source (Sannicandro, 2019). For producer Thessa Torsin (aka upsammy), mixing field-recorded acoustic and electronic sound sources in tracks is a way to contrast the details and moods of the organic and the digital by *zooming in* on them:

> With my Zoom recorder I recorded some field recordings and percussive sounds, connecting the music to the visual ideas. I wanted to look for contrasts sound-wise, make it sound organic yet digital. Using soft melodies but more bright and icy percussion for example. The act of 'zooming in' was a way of working throughout the whole process: looking for details in sound and researching my surroundings closely to find those sounds and moods.
>
> *(Cain, 2020)*

Producer Alexander Dennis (aka EPROM) makes field recordings of the subtle noise elements of his studio's equipment. Dennis uses samples of tape noise, vinyl crackle, reverbs, and noise floor sound to add an almost subliminal sense of decay and shadow to his music through what he evocatively calls a *softly breathing field of noise*:

> Almost every sound is decayed, or shadowed, or manipulated in some way. I've been interested in injecting noise into my productions lately; using a lot of tape noise samples, distortion, old spring reverbs, bits of vinyl crackle, and the noise floor that my synths generate to create this softly breathing field of noise.
>
> *(Marlow, 2013)*

Sound artist and producer Joseph Kamaru (aka KMRU) uses field recordings as samples, as input for sampler instruments, and as textures for sound synthesis. Kamaru's workflow often involves listening through his recordings in search of "sounds/patterns that can be building blocks for a full piece and build off of that" (Housego, 2020). Recalling Geir Jenssen's using samples as subliminal atmospheres, Kamaru will sometimes use a single lengthy soundscape as an environmental drone upon which he improvises with a synthesizer.[12]

For both creative and copyright reasons, producers often heavily process field recordings so as to make them unrecognizable. This can lead to an interesting duality: as the recording's original source becomes less and less important, its processed sound becomes a one-of-a-kind beauty. For producer, artist, and writer Richard Skelton, recordings of acoustic instruments can be transformed using any number of techniques to reveal hitherto unheard ideas. Here, field recording is the first step of a composing process intended to generate a music that's more than the sum of its parts:

> I record acoustic instruments, and the challenge is to record them as authentically as possible...Once recorded, the process of composition involves transforming the sounds whilst somehow retaining something of their original essence. For me composition involves trying to coax 'music' into being from these disparate recorded sounds. It's a bit like making a sourdough starter. You add the constituent elements, stir, and then wait. Hopefully, if the conditions are right, something will happen and then it's essentially out of my hands. Something has occurred that is more than the sum of its parts. From then on the challenge is not to try and exert my own will over it. I might help give it shape here or there, but for me music is 'other' and deserves to be treated with awe.
>
> *(Fischer, n.d.)*

Finally, producers can learn from the practice of sound designers who work alongside film composers. For example, recordist Robert Dudzic emphasizes that ambience speaks for itself—it doesn't require effects processing to situate a listener in a realistic space because it embodies that space:

> When you record something like ambience, and you add anything to the sound, you are taking away from the originality of that sound. Don't buy the latest plugin that's promising to make you sound bigger and better. Spend your money on a nice handheld recorder, or a decent microphone. Because people close their eyes and want to feel it, they want to believe it. They want to feel like there are there.
>
> *(Stout, 2018)*

Veteran Hollywood sound editor Mark Mangini is also an ambience proponent who makes field recordings for adding perceptual complexity to a soundtrack. In a documentary about the film *Dune*, Mangini sources soundscapes in the Mojave desert, where he records sand sounds at close proximity. Echoing Villalobos' and Hopkins' views of the acoustic world's vividness and the futility of replicating it artificially, Mangini explains that the basis for a film's sound design is the complex acoustic environment of real life:

> We can be more successful in our sound design when we start it with acoustic recordings. The reason for that, I think...is that subconsciously there are embedded

> elements of acoustic recordings that tell the brain, *This is real.* And it's all about the time arrival to the ear and the acoustic environment that a sound lives within that synthesizers or electronic sounds don't inherently have. You can add them in post-production by adding reverbs and delays and things like that, but they're never as complex or as rich as the acoustics that you get in real life.
>
> *(Warner Bros., 2022)*

Making drones

Drones are a controlled sonic stasis, often limited to just one or two continuous pitches, usually a tonic note and the fifth and octave above it. These pitches create a sense of space and depth without requiring the literal, full-frequency spectrum space of a field recording or a lush reverb modeled on the impulse responses of an acoustic place. In this way, drones are a virtual ambience.

Because drones typically do not change their pitches over time, they're used in music to create a sense of endlessness that functions as a background texture for solo vocal or instrumental parts. For example, in North Indian classical music a drone's notes are plucked on the *tanpura*, a long-necked 4-metal-stringed instrument (Figure 7.2). The *tanpura* player plucks the first (*sa*) and fifth (*pa*) degrees of the music's *raga* (broadly,

FIGURE 7.2 The *tanpura*, an ambience instrument

scale). The plucks are repeated in slow alternation to produce a drone that's "a resonance rather than two separate standalone notes" (Chaudhuri, 2021: 114). The *tanpura*'s curved bridge design creates another sound layer in the form of a subtle buzzing that's rich with harmonics.[13] Against this resonant backdrop the vocalist or instrumentalist improvises melodies in the raga. During the performance, the drone is background but we notice it here and there—in the spaces between the soloist's melodic phrases, in moments of dissonance where the soloist's pitch moves a half-step near the drone, and in moments of euphony when the soloist and drone emphatically convene on the same pitch.

Electronic music producers use drones in a way not unlike its backdrop functionality in Indian classical music. Two differences, however, are that producers create timbrally unique drones out of a track's sounds and processing effects, and then use these drones to seed new compositions. Sometimes referred to by producers as *harmonic landscapes* or *ambient beds*, drones are tools for revealing the tonal possibilities of one's musical material.

Producers typically make their drones by turning an element of the music into a continuous or looping sound. There are numerous ways to do this, but three methods are common. The first is to saturate an element in a long reverb—100 percent ambience—so that it blurs into an amorphous cloud. A second way is to put an element through a granular synthesis effect that splits a sound into grains whose speed and length can be adjusted to turn a short sound into a continuous one. A third way to make a drone is to dramatically slow down a recording using software such as PaulXStretch (PaulXStretch n.d.). PaulXStretch'd audio can be made to unfold incredibly slowly, so that a melody or chord progression becomes a gargantuan hum.[14] Producers use elements of each of these techniques—adding reverb, granulating, slowing down audio—to create endless varieties of drone.

But why make a drone? One reason is that it's an ideal form of recursive music production, a topic explored in Chapter 8, whereby a producer reuses what is already at hand, folding material back upon itself via processing to create a new part. Another reason to make a drone is to inspire close listening to its audio from which a track could be built. For example, producer Fred Gibson (aka Fred again..) hears drones as framing the emotional hues of the vocal samples he uses in his tracks. Gibson will create multiple drones from a single sample to generate a *harmonic landscape* for melodies:

> Long drones—[are] different ways of framing what the person is saying. Generally with samples I'll make like five or six drones that it can live in, and the harmonic landscape of those drones will affect the emotion of the song—that's what they're born out of.
>
> *(Lowe, 2022)*

Making a drone also jumpstarts a project with a set of pitches that establish a key or chord center. The producer Nicholaas Douwma (aka Sub Focus) makes drones by processing his live playing with reverbs and PaulXStretch:

> This tends to be something I start with a lot: I make drones in the key of the song that I'm working in. The way I normally do them is that I record a long pass of me playing an instrument. There's one I made from a bleepy square wave patch in Serum [synthesizer] which is put though Valhalla Shimmer [reverb], so it's got a long

> reverb on it. There's another one I've done with a string [sound]: I'll record a long pass of overlapping notes and then I put that into PaulXStretch."
>
> *(Tape Notes, 2023)*

Like the *tanpura* in Indian music, drones generate a tonal texture over which other parts can flourish. For Douwma, drones replace silence with tone. This gives him something to play off of:

> It's been a nice technique I've used early on in projects, because it's nice not making music from complete silence. Normally the next step of my writing process is to write a bass line or figure out what the root notes of the chords should be. It's good to have something to do that *to*, otherwise you're just doing it in isolation. So I quite often have these [ambient] beds, and also quite often re-use them from other songs.
>
> *(Tapes Notes, 2023)*

In sum, making drones is the producer's way of unifying a track by transforming its elements into harmonic landscapes and ambient beds. Ambience isn't the same as ambient, and one doesn't have to be making ambient music to experience the ear-opening practice of making drones. Turning a part into a drone reminds the producer of the malleability of sound in music production. Try making a drone to transform a track's elements into creative resources, revealing once again the potential of sounds within our musical system to be alive at many levels at once.

Effects are all about affect: effect chains

In a DAW, effects are a compelling point of production focus because there are so many of them and they can be endlessly combined. For the producer, effects take conventional sounds into new and often strange territories by adding traces, echoes, ambiences, textures, harmonies, sub-tones, presences, and implied rhythms. One can overdo effects and drown a sound source with them, or use effects lightly so that they're felt as much as heard. Effects are a producer's magic wand, capable of morphing this sound into that one and conjuring an enchanting aura out of the humblest audio source, reminding us that a track's sonic profile is always fluid, a feeling in progress. In short, effects are explorations into timbral-textural re-shaping, and are ultimately *all about affect*. Referring to the feel of reverb and delay plug-ins by the software company ValhallaDSP, producer James Blake describes their effects as emotional buttons:

> The Valhalla plug ins are pretty unbelievable. All these things, they start as plug ins, but they turn into emotional buttons that you can kind of press and go, *Ok, I want it to feel like this*. That's kind of how I feel about plug ins.
>
> *(Tape Notes, 2023a)*

From simple processing of instrument timbres to complex, multi-timbral transformations of a mix, effects are affects and an effect is only as useful as it is affecting.

One way for the producer to learn about effects is to use them in various combinations often referred to as *effect chains*. Effect chains are routes for sounds to travel, flowing from one effect into another and then another, picking up distortions, resonances, and colorations along the way. They're sound processing black boxes the producer assembles without necessarily knowing what their composite sound will be until she hears it. Effect chains allow the producer to work like an alchemist, to "unveil a continuous incantation of magic spells" (Bach, 2003: 6) in order to achieve the grand results that come when sounds are free to metamorphosize. "Effects" notes Kodwo Eshun, "are now acoustic prosthetics, audio extensions, sonic destratifiers, electric mutators, multipliers and mutagents" (Eshun, 1999: 006). Keep tweaking the effects in your chains as well as their parameters—more of this, less of that—until you hear the newly interesting.

In tandem with his systems thinking, Brian Eno was an early proponent of effect chains. He used the term *treatments* to describe connecting several conventional effects such as chorus, reverb, and delay to make new timbres out of whatever sound he fed through the chain. Harold Budd, a frequent collaborator of Eno's in the 1970s and 80s, explains that these treatments were more than mere enhancements. They were new sounds unto themselves:

> Brian demonstrated to me this brand new world of making solo music that was hardly solo. It was employing chorus effects and delay effects to alter music in a way that they weren't designed for. We were using them at the extreme left end of the spectrum, where you're hearing the effect as the sound itself, rather than using them to industry standards whereby they enhance a fully created sound. I just put a microphone on the piano and this is going through a configuration of processing machines, then into the mixer and onto tape.
>
> *(Prendergast, 1986)*

I'm playing a piano sound on a tiny keyboard controller that isn't ideal for improvising because my hands can't flow on small keys. When a producer is frustrated by controller ergonomics, a solution is to try to do more with less—as producer and harpist Nala Sinephro puts it, to pursue "the possibility of pressing one note with a shit-ton of intention" (Monroe, 2022). Instead of trying to play a lot of notes, I alter the piano sound with an effect chain to try to make one note interesting on its own. Focusing on the enchanting details brought forth by the effects, soon I've forgotten about the limitations of the tiny keyboard because one note is triggering a huge sound.

I've gone into my library of effect chains and randomly auditioned a few to re-familiarize myself with sound design work from past projects. Many producers use effect chains this way. Producer Thomas Nicol (aka Tom VR) explains that "I have loads of saved effects chains that I use constantly, created by myself, so I always think about these like my main plugins in a way" (Mullen, 2022b). Producer Nathan Micay saves chains the same way, even though, as we learned in Chapter 2, he doesn't think of himself as a sound designer who builds sounds from scratch:

> I'll take a preset from Diva [synthesizer] and I'll usually process it or I'll make some chain. I like building chains from the ground up, I don't like building sounds from the ground up. Although I guess you could argue that a chain creates a new sound anyway.
>
> *(Mullen, 2023a)*

Effect chains *do* create new sounds. Producer Jon Hopkins builds very long chains which he later re-uses on sounds they weren't originally built for, just to hear the result:

> I'll do things like have a long chain of plugins I have saved in Ableton that have been built to process one particular sound, but then I'll try putting different sounds into that specific, really strange chain of plugins and see what you end up with…I'll have plugin chains of maybe 15 plugins running into each other and the results end up so far away from what you started with. Why not try playing a piano or singing into that to see what happens?"
>
> *(Mackintosh, 2021)*

An effect chain is two or more signal processors connected serially (or in parallel) so that a sound passes through, and is altered by, each of them. A simple example of this is a reverb, which situates a sound in a space, and a distortion, which adds saturation and grit. Each of these effects has a wet/dry knob that controls how much of the effect is applied to the sound source running through it. One can arrive at appealing results when each effect's wet/dry knob is set to less than 50 percent, so that the effects tint the source instead of engulfing it.

Producer and guitarist Leo Abrahams thinks of his effect racks in Ableton as "virtual pedal boards." Abrahams splits his guitar's signal through multiple effect chains running in parallel (instead of serially) so that "through that simple starting point, you can generate quite complex but interactive layers of sound while being in full control of that space" (Turner, 2024a). Abrahams describes his effect racks/virtual pedal boards as a kind of musical system, the topic we explored in Chapter 2. This system functions as a collaborator that generates serendipities and fosters discovery

> because I could construct effects chains that reacted in slightly unexpected ways to whatever I was putting in…The beauty of this system is that it helps me to forget what I'm doing. It feels like an opening to curiosity and discovery, so it's not that I want the computer to help me to write music, it's more to do with creating an environment where music can happen.
>
> *(Turner, 2024a)*

A more complex chain of effects can simply have multiple instances of the same effect, each one doing different things. For example, one might use multiple delays, each conjuring different spaces. Producers Ólafur Arnalds and Janue Rasmussen use this technique to sculpt the techno they make as Kiasmos:

> You've got maybe two [Roland] Space Echos, or if you've only got one you just run it twice, and you really drive it, you have a lot of feedback and maybe even change the time a bit so it's kind of wobbly. Then you cut all the low frequencies all the way up to 4kHz, so it's just the really high silky stuff that's left. That's a really wonderful, high-end ambience that just fits on top of anything. We've used that quite a lot.
>
> *(Mullen, 2024)*

In addition to the number of effects, the order of effects in a chain also impacts the chain's sound. For example, a distortion before a reverb creates a distorted sound dispersed in a space, while a reverb before a distortion creates a space compressed into distortion. One might also place a compressor at the end of a chain to amplify details of its effects. Experiment to explore permutations of your effect chains.

While there are no rules as to how to make effect chains, we might do well to follow Eno's *economy is elegance* dictum discussed in Chapter 4: simple chains can lead to complex effects. The simplest chains, like reverb + distortion, can be sound-warping beasts. Moreover, the accidental discoveries we make while experimenting with chains inspire us to iterate upon those effect combinations that produce the best results. A rule of thumb then, is *keep the same effects but change their parameters*. Lastly, making chains occasionally leads us to notice *a sound that sounds like a potential musical idea*. The production lesson here is to not expect musical ideas to present themselves before you've begun. Instead, we work our way towards ideas, or sneak up on them while doing something else, like tinkering with an effect chain that turns one note into a huge sound. This is *economy is elegance* in action.

In sum, working with effect chains is part of what makes composing music in a DAW fundamentally different from jamming on an instrument or writing notation. In their open-ended unpredictability, effect chains are portals to new sound worlds. Building chains through an iterative, trial-and-error workflow will exponentially extend your sonic imagination as you turn conventional sounds into ones you've never heard before.

Notes

1 For a history of Muzak, see Lanza (2004).

2 The first World Soundscape Project recording was *The Vancouver Soundscape* (Truax, Westerkamp, and Schafer, 1973). This recording featured the sounds of Vancouver's soundscapes, recorded between September 1972 and August 1973, along with Schafer's spoken commentary on the city's acoustic design.

3 See, for example, Truax (1978; 1996) and Westerkamp (1974).

4 In *The Soundscape,* Schafer presciently observed that the "blurring of the edges between music and environmental sounds may eventually prove to be the most striking feature of all twentieth century music" (Schafer, 1977: 111).

5 For example, Steven Feld and Chris Watson have released recordings of acoustic environments including rain forests, bells, and glaciers. See Feld's *Voices Of The Rainforest* (1991), *The Time of Bells, 1* (2004), and Watson's *Weather Report* (2004).

6 See, for example, Dan Gibson's *Solitudes: Exploring Nature With Music* (Gibson, 1993) among many other similar releases.

7 A repository of ambience are the immersive ASMR YouTube videos that digitally render "cozy" libraries and home interiors with a soundtrack of a crackling fireplace and rain, thunder, and wind sounds. For example, see New Bliss's "Royal Library" video (New Bliss, 2019). Intended for relaxation or productivity, such videos reframe Schafer's soundscape concept as unobtrusive soundtrack. These are the most ambiently ambient music because you can imagine in them whatever you want.

8 For example, Alan Lomax's numerous 20th century field recordings of folk and blues musicians in the United States and Europe.

9 Sound designers working on the *Dune* films used a similar approach, playing synthesizer parts through physical resonators made of wood and metal (such as La Voix du Luthier's *Onde*) and recording their sound in a room. This process "preserves the acoustic and organic aspect of the sound, so you don't think it's a synth" (Inglis, 2024).

10 An example of capturing the spatiality of acoustic music recordings is Villalobos' and Max Loderbauer's *Re: ECM* (2021), which samples instrument sounds, voices, and resonant spaces from ECM's recording catalog. Listen to the evocative track "Reblazhenstva."

11 In the 1950s, record producer Sam Phillips was light years ahead in thinking about how to capture sonic space. Phillips thoroughly explored the ambiance options of his Sun Studio, for example using the studio's "secretary's office as an echo chamber" (Blitz, 2016).
12 A documentary on KMRU shows him playing along to one of his field recordings using Soma Laboratory's Lyra-8 drone synthesizer (Em, 2020).
13 The *tanpura*'s drone sound, notes Amit Chaudhuri, "may be the cultural forefather of trance music" (Chaudhuri, 2021: 144).
14 There are many PaulXStretch'ed drone sound examples on YouTube, among them a version of A-Ha's 1980s pop hit, "Take On Me" slowed down by *800 percent* (A-Ha, 2011). At this speed, the song liquidizes into slow-moving waves of harmonies.

8

TOWARDS DYNAMIC STRUCTURES

Introduction

Electronic music producers' tracks play with a tension—between repetition and change, stasis and development, staying in this moment and building anticipation for what comes next. No matter what style of music we make, our omnimusical production practices involve numerous decisions about how to unfold a track over time as a series of simultaneous and successive sonic events in "a now whose content changes ceaselessly" (Burrows, 1997: 529). An essential skill is knowing how to configure music so that its elements change ceaselessly yet meaningfully, so that sounds create the sense of going somewhere. At some point in a workflow we ask ourselves, *How can I grow this track, build out its parts, and develop its ideas?* The compositional problem, as the producer Simon Green (aka Bonobo) puts it, is essentially how do we turn "eight bar loops forever...into a cohesive track"? (Tape Notes, 2024). This chapter explores this problem through the topics of loops, grid musical thinking, tonal evolutions, form, and ways of developing ideas including recursive production, scaling, prolation canons, and transitions. But first a note about dynamic structures.

In Chapter 2 we borrowed a concept from architect Christopher Alexander to understand electronic music production as a way of creating a sonic world alive at many levels at once. This concept, which reappears throughout this book, describes the way parts in a track ideally interact with and respond to one another. But it's also an approach to conceptualizing the structure of one's music. For Alexander, nature's structures feel harmonious or aesthetically pleasing in some way. We notice nature's structures, for example, when we take a walk and sense the "dazzling geometrical coherence" around us.[1] Nature's structures are achieved *dynamically* through what Alexander calls continuous very-small-feedback-loop adaptations (Brand, 1995: 21). The applications and meanings of these adaptations in the context of music production are the subject of this chapter. As we'll learn, producers find ways of making the elements in their music interact with and adapt to one another. This creates a *dynamic structure* which is an arrangement built upon calibrated change, development, and adaptations on multiple levels of musical action—from tiny sounds to long phrases, from

DOI: 10.4324/9781003585145-10

sections to entire tracks. It's this dynamism at different resolutions that gives a track geometrical coherence and therefore makes it feel harmonious, in tune with aesthetic qualities we value. Let's begin by considering two ways producers achieve dynamic structures: ways of loops and ways of tonal evolution.

Ways of loops

The breakbeat, the sequence, the riff, the inherent rhythm, the timeline, the *tala*—a loop is a groove propeller, a repeating bit that captures a vibe and spins it around and around until its sound-feeling is airborne, aloft on its own motion, soaring over time. A loop proposes being circular for a while, as if saying *here is a way of going as a place of being.* Loops are patterns repeated, played by musician or machine, loose or quantized, going around until trance states beckon. The musicians smile and drop metaphors because the rhythm is *deep, in the pocket, swinging.* Now you hear details in the bit, depth in its changing same. Loops amplify attention, attuning us to notice small differences that reveal repetition as transformation.

Loops in electronic music production take us back to the *musique concrète* experiments of Pierre Schaeffer, whose *Étude aux chemins de fer* had a groove to it. But as discussed in Chapter 2, it was primarily drum machines, sequencers, and samplers that brought loops to music's foreground. From disco to techno, from rock to hip hop, every style of 20th and 21st century popular music has been touched by the loop. Today, using loops is the quickest production route to a dynamic musical structure with a trance quality. Let's consider some ways producers have worked with them.

To make *Thursday Afternoon, Discreet Music,* and *Music for Airports,* Brian Eno innovated a practice of recording long sequences on tape recorders and then cutting them into loops of different lengths. Each sequence's loop point was unique, creating what Eno calls an "unlocked" music:

> "When I make loops on a sequencer, I always try to play them all the way through, so I play the whole part, then I listen to it, and quite often I find a long section that I like. Loop that, cut it up so that the loop doesn't recur regularly. The idea of always editing in straight vertical cuts is the most single annoying thing about most of that music. Because a whole part of my feeling has been to make music that is 'unlocked.' And all that stuff like *Thursday Afternoon, Discreet Music* and so on, is very deliberately that: music where the elements float separately from one another."
>
> *(Schütze, 1995)*

Producer Simon Green (aka Bonobo) uses a similar, *make the elements float separately from one another* technique to make dance music. Green avoids having parts repeat at the same points, because doing so overemphasizes dance music's default 4/4 metrical feel. Like Eno, his solution is to unlock repeating parts by making them of varying lengths. This way it takes several bars for the loops to re-align. Green used this technique to make his 2024 track, "Expander":

> I'll do something where I'll try not to loop everything in fours. So the hi-hats will be a sequence of five notes or a bar and a half, and then the snares will repeat in threes.

> Every hit is slightly different; you get a full sequence of 16 bars before anything is exactly the same, even though it's very subtle. It kind of makes it feel a little looser and you're not really getting that repetition as much because things aren't hitting in exactly the same place each time.
>
> *(Tape Notes, 2024)*

Graham Massey, of the quartet 808 State, explains how in the late 1980s the group composed by recording dozens of loops and then arranged them into a finished track by bringing elements in and out of the mix. 808 State also worked with loops of varied lengths. Their low-tech tool for playing the loops was the Mute button on their mixing console:

> The way we used to compose was sort of like filling a 24-track tape full of loops, basically. Some of them will be 8-bar loops, some of them will be 2-bar loops, some of them will be 6-bar loops. So they'd all spiral 'round each other while you composed on the Mute buttons. And one of the aspects of there being four of us in the band—is eight hands operating the mixing desk. Because we didn't use automated mixing back then.
>
> *(SOS Electronic Music podcast, 2023)*

Producer Brian Leeds (aka Huerco S.) makes music in which "textures and rhythmic artifacts interweave with one another" (Shopmaker, 2022). Leeds achieves this interweaving by spending a lot of time listening to his loops. Thinking like a sculptor, he likens loop-making and listening to their repetition to conjuring a physical sonic object, "building these patterns and loops and kind of re-listening over and over again". Leeds' process is itself on loop, "almost like sculpture or something, just adding and subtracting things" he says (Shopmaker, 2022).

Producer Dan Deacon considers loops the core of his workflow. Whether combing through recordings of live musicians or working alone with electronics, Deacon always seeks the loop: "I'll focus in on a section. Either if I'm looking for a sound, or if I've recorded a long session with a player, or if I'm just building something from scratch. It revolves around a loop." Once he has a loop, he finesses it as he listens to it repeating, "making slight changes to various aspects of the sound". Decon's aim is to create "micro-fluctuations, and modulations and changes" within the loop to make it a dynamic structure alive at many levels at once (Webb, 2020).

Some producers find loop opportunities in production gear. For example, Jan Jelinek used an Ensoniq ASR-10 sampler to make his aptly-titled 2001 recording, *Loop-Finding-Jazz-Records*. Jelinek's attention alighted on the ASR-10's "Auto Loop Finding" parameter which allows a user to set very short loop lengths (Ensoniq, 1992). These short loops can generate interesting sounds. Jelinek explains that the best loops are the synergistic product of multiple sounds combining and cooperating to make a new composite:

> Creating loops is based on synergy, like a chemical reaction. It can happen that two rather uninteresting audio samples start shining only in combination. A beautiful loop is mostly based on cooperation of different sound sources. That's why such work is still exciting, even after such a long time. I never know how the particular components will react in advance.
>
> *(Fosl, 2018)*

Once Jelinek has a loop, he lets it repeat and listens to assess its subtle perceptual work. Ideally, a loop eventually becomes an almost imperceptible, neutral background, not unlike the drone of the Indian *tanpura*, discussed in Chapter 7:

> Usually I'm creating repetitive loops which stream sound in a continuous mode. If it is successful, it reaches a state of limbo: you rather don't notice the music anymore, and at the same time, it expands the room with an acoustic, becomes a part of the environment. You can call it a neutralization: it is the point where music shifts into function.
>
> *(Fosl, 2018)*

Producer Ben Edwards (aka Benga) has a longtime interest in analog synthesis. His 2008 recording *Twenty Systems* is an archeology of iconic synthesizers that traces a history of various models from the late 1960s to the late 1980s. Each track features one instrument recorded without effects to document the souls of vintage machines. To create the music Edwards used a single loop of melodic sequences to trigger the synthesizers' various sounds:

> In a way I wanted to take a snapshot of the sound of each synthesizer and just see how it would sound if I applied the same principal to each track, which was to set up a loop of sequences and let the track evolve naturally rather than impose a kind of idea on the actual composition. They are all quite similar in a way because that is the simple idea I had for each track.
>
> *(Doran, 2008)*

Producer Carmen Villain creates loops to have something to play along with and inspire new ideas:

> I'm always looking for interesting loops. Just anything that might trigger some sort of idea. I find it really difficult to just sit down and start sort of composing something from nothing. I've done it before, but I find it much more inspiring to have something to jam along with. Because that might actually give me more ideas when it comes to harmonics or melodies or something that might actually just spring out of sound.
>
> *(McTernan, 2023)*

For producer Will Long (aka Celer), loops are the foundation of his slow ambient music. He makes loops by playing parts on acoustic instruments, recording the results, and then mixing the parts together to make a composite. Like Eno and Green, Long eschews loops that obviously repeat. Describing how he made his 2021 recording, *In Light Of Blues,* he explains a process of

> playing things, then matching and mixing them together, in order to make something that sounds like it would make a good loop. But in this case, I didn't loop it. I combined different elements that sound similar and made them more singular, only being heard once, or repeated with slight variation after a few minutes. It's built with blocks of 'loops' that don't repeat.
>
> *(Allen, 2021)*

Producer Daniel Snaith (aka Caribou) makes loops to have source material for future tracks, a practice that connects to Chapter 1's discussion of the musical index card. "I'll make two or three 30-second long loops and they just accumulate" he says. "Every day I make a couple of ideas and then they pile up, and I kind of sieve through them to find the ones that turn into tracks" (Future Music, 2020).

Producer Axel Willner (aka The Field) builds his music entirely out of microscopic samples of other songs. To make his recording *From Here We Go Sublime* (2007) his workflow involved "looking for bits and pieces in it that I really like and I'm trying to rearrange it" (Pitchfork, 2007). For Willner, a track's structure can emerge out of "the perfect loop, the loop that just sits perfectly" (Gonsher, 2013). Ideally, the repetition embodied by a perfect loop is a music's structure— in other words, *the loop is the music*. In a 2013 interview, Willner recalls listening to SND, the duo of Mark Fell and Mat Steel. It was while repeatedly listening to the "locked grooves" of SND tracks that Willner heard different rhythms emerge as a by-product of his mind wandering. From this, he reflected on the perceptual power of loops in his own music:

> There aren't necessarily things happening in my music all the time, and it's always interesting to hear how people experience it in a totally different way from me...If you sit and listen to a sound for a long time you don't need anything else to transpose it or change it, you just alter the experience slightly with your mind.
>
> *(Gonsher, 2013)*

Our final example of using loops to structure music illustrates how loops don't always survive linear arrangements. Explaining how he made his recording *Decay* (2014), producer Phillip Sollmann (aka Efdemin) observes how a loop that works beautifully on its own can fail in a track:

> It is an interesting phenomenon that you can listen to some loops for hours, but when you start arranging them into a time-frame, they will crumble and run through your fingers like sand. On *Decay*, there are various examples where it was simply more effective to present an almost static arrangement and allow developments to take place on a micro-level and within the sounds.
>
> *(Fischer, 2012a)*

Grid musical thinking

The omnipresence of loops in electronic music production reflects a shift in music practices from the mid-20th century to today, from linear through line-based composing (inherited from the European classical tradition) to cyclical, loop-based composing. This shift was the result of the profound influence of African, African American, and Caribbean rhythm aesthetics on popular music worldwide. What is fundamental to all of today's electronic dance musics are kinetic beats and the sense of vital, forward propulsion they provide. The jazz drummer's ride cymbal pattern, EDM's four-on-the-floor kick, bell and clave timelines, drum machine beats—Black rhythm aesthetics are music's force multipliers around the world of music.[2]

FIGURE 8.1 The grid matrix on Ableton's Push controller

Loops and Black rhythm aesthetics connect to the grid musical thinking in electronic music production. A quick history: in 1988, the Japanese electronics company Akai collaborated with Roger Linn on the MPC60, a sampler-sequencer-drum machine that was the first grid-based MIDI instrument. The MPC had 16 rubber pads, arranged in a 4x4 grid matrix, for triggering samples and sequencing beats. Linn's idea for the grid interface came from watching Stevie Wonder use his earlier drums machines and realizing the need for a more tactile-friendly design. Since Linn's MPC the grid MIDI controller (Figure 8.1) has become a preeminent symbol of music production. The grid form presents African music's cyclical timeline concept as a matrix, a series of squares within a larger square. The grid visually represents 4/4-time, 16th note beat subdivisions, 4-bar rhythm patterns and step-sequencing, quantization, and the triggering loops method of composing. Like the turntable and the steel pan, the grid controller is one of the few genuinely new musical instruments of our time.

The grid controller reminds us of the parallels between musical practices and art practices. Case in point: in the 20th century, grid forms were an emblem of modernist visual art. Grids organized many seminal paintings and sculptures, such as Piet Mondrian's *Composition with Grid #1* (1918), Jasper Johns' *Grey Numbers* (1958), Agnes Martin's *Untitled* (1965), and Sol LeWitt's *Color Grids* (1980). What attracted visual artists to grids? Maybe it was a sense of precision and order, non-representation, and repetition. In her 1979 essay *Grids*, critic Rosalind Kraus explained how the grid as an aesthetic form is "flattened, geometricized, ordered, it is antinatural, antimimetic, antireal. It is what art looks like when it turns its back on nature" (Kraus, 1979: 50). The grid, Kraus said, is "a mode of repetition" whose form represents "the antinarrative" (Kraus, 1979: 61, 64). More speculatively, perhaps the modernist art world's explorations of grids was a visual parallel to the loop-based, Black rhythm aesthetics that were reshaping 20th century popular music.

Grid art may have also influenced the direction of minimalist composers who, in the 1970s, began substituting strictly repeating and emotionally neutral musical forms for classical music's expressive complexities. For example, Steve Reich and Philip Glass adapted rhythmic concepts from African and Indian musics to use as structuring devices: Reich used timelines and polyrhythms, while Glass used *tala* (rhythmic cycles), melodic cells, and additive rhythms. Interestingly, all these techniques are grid-esque. Coming full circle, the minimalist sound would eventually influence a generation of music producers and DJs who were building upon Black rhythm aesthetics through their electronically-generated beats.[3] From loops to repeating geometries in both visual art and musical practices, the grid form connects across artistic milieus.

In sum, grid structures, grid devices, and grid thinking are omnipresent in electronic music production today. But while producers work with and within grids, they also seek ways to go *off the grid,* especially by incorporating little timing discrepancies. Back to Simon Green (aka Bonodo), who explains a practice of rhythmic finessing now common among producers. "I'll go through making micro-adjustments on every hit—on the velocity, on the position, on the quantize, just to pull everything off the grid" he says (Ableton, 2017a). So if Kraus was right about grid forms in the art world being *flattened* and *antinatural,* then perhaps what producers are doing by loosening grooves with micro-adjustments is finding yet another way of making their music more 3D and natural, alive at many levels at once.

Ways of tonal evolutions

A second way of achieving dynamic structures is to construct music that enacts evolving shifts of notes, using chords that continually morph into new constellations to engage the listener. Such *tonal evolutions* create active sonic textures rather than passive ones, their trajectories over time leaving traces of their consonance and dissonance filigree.

A way to create tonal evolutions is to compose/improvise an initial chord sequence or melody and then use this to *seed* other musical materials—from new chords and melodies to textures, rhythms, and even drones. As discussed in Chapter 2, there are many kinds of uncertainty built into chord or melody improvising. This uncertainty is creative gold. Our improvising poses questions and then answers them:

How long will I play for? Four bars or sixty, 30 seconds or 5 minutes?
What will the tempo be?
Will I repeat motives I discover along the way, or continue onto ever-new phrases?
Will I leave space between the notes, or pursue denser constructions?
Will I stay within a small or large dynamic range?
Will I leave my mistakes in, or correct them after the fact?

In other words, our chords and melodies captured as MIDI or audio is ample material to explore. Consider six ways of deriving tonal evolutions from an improvisation.

Identify the good bits. Somewhere in your performance there may be a gem that stands out. It could be a clear idea in the middle of a long phrase, or a beautiful cadence at the end of an unremarkable meander. Identify the good bits so you can evolve them further.

Consider the structure of the track-to-be. The unique ebb and flow of your performance can organize a track. Your playing may have implicitly delineated sections, and therefore a proto-form, that you can use. Large-scale structure is inherent in small-scale form.

Derive other parts. Since you can use MIDI to trigger as many sounds as you have instruments, there are chords, bass lines, and little subliminal sounds latent in your improvisation. To excavate these parts, copy your MIDI to other tracks to play other sounds. A chord sequence becomes a bass when you pitch it down and mute all but one of each chord's notes, or the sequence's highest notes solo'd become a melody.

Generate permutations. Use your DAW's permutation-making tools to play with your performance's MIDI: re-pitch, repeat, mute, invert, stretch, compress. Practice permutation-making until it's an inevitable part of your workflow.[4]

Bounce to audio. Process MIDI and then resample it to an audio track to bring the sound you began with far from its starting point.

Use bounced-to-audio to generate new MIDI. A bounced-to-audio performance can in turn generate a new MIDI part. As we learned in Chapter 2, audio-to-MIDI conversion is imperfect because the DAW sometimes hears phantom pitches that aren't there. Use this mishearing to seed more sounds.

Form

Form is how a music unfolds over time, a sonic schematic that specifies what goes where, entrances and exits, and what happens now. Decisions of form are guided by the fact that in music, as in a novel (but unlike a painting), everything can't happen all at once. A music's form is its structure in motion. Sometimes form is predictable, the result of repeating loops and sequences arranged like Lego blocks—with fills and transitions every so many bars to cue the listener that something is about to happen. Predictable form is a by-product of predictable cut-and-paste arrangement decisions facilitated by using the DAW in predictable ways. But producers have endless options for organizing their musics into unique forms that feel dynamic.

In many styles of instrumental electronic music, form is modeled on songwriting's verse-chorus-bridge designs. In the arrangement view of a DAW, the producer positions repeating loops and sequences into a linear array of intros, main themes and variations, build ups, breakdowns, drops, and outros, appearing in 8-, 16-, and 32-bar increments. This time-tested "'geometric' approach is useful for adding interest and complexity to a mix within a logical system" and helps the music feel like it's going somewhere (Trimbacher et al., 2024: 19). Paul Hartnoll, of the duo Orbital, recalls structuring tracks in terms of symmetrical ABA forms, where the B sections are simply variations in mood:

> In terms of track arrangement I tended to favor A, B, A...You start with something, you go into something, then you come back to what you started with, but make it a bigger version. Or, I quite like the Part Two where you do something, and then do a kind of moodier second part.
>
> *(Future Music, 2022)*

The ABA form is a conventional structure that can anchor a producer's abstract sound design within a logical system. Producer Lewis Roberts (aka Koreless) explains its utility:

> Sometimes if what you're working with is already quite abstract, it can be really useful to just try and force it into a very conventional structure, either within it, within a track, or within a whole album or something. It can be a nice way of making abstract stuff make sense—to force it into a very rigid and straightforward structure.
>
> *(Roberts, 2021)*

Similarly, producer Joe Moynihan (aka Yamaneko) applies dance music's geometric structures to ambient tracks:

> Some of the things I love most about grime—like eight bar structures, big low frequencies, very sparse arrangements—I'll make an eight bar ambient tune and try to channel that energy into a tune that you could listen to while doing yoga. It still feels earnest. It's just translating my love of grime, eight bar or whatever, into a context that I feel quite comfortable working in.
>
> *(Murray, 2018)*

A quick case study in form is "Goodbye", a 2022 track by veteran producer Mike Paradinas (aka µ-Ziq), which uses 8- and 16-bar sections without transition markers (e.g. fills) between them. "Goodbye" is a kind of changing same made from a few parts appearing and disappearing in various combinations. Each section is similar yet unique, arranged in a seamless sequence, whose orchestration keeps shifting. In sum, compelling musical form is often simple, geometric, and unhidden.

Be simple, but don't be *predictable.* Adventurous producers understand form to be an ambiguous question in search of a structural answer. Ideally, when you're making music, you don't know where you're going form-wise because you discover it as you go along. This happens most clearly in improvising, but also when sound-designing, mixing, and arranging. For example, you might play something you never planned on playing, create a sound you never planned on creating, or find an orchestration of parts you didn't foresee. Such accidents of discovery and serendipity generators often lead us to new forms. The more we make music the more we realize that answers to questions of form are different for every track. We find form as we go, sometimes starting a project in what feels like a state of chaos and gradually moving towards order. Producer Jon Hopkins shares a lesson from the approach he and his musical collaborators used to evolve the form of *Ritual* (2024), a 40-minute, multi-movement work. Hopkins advises: "Sketch out the big picture first, just work as quickly as possible, dropping things in [the arrangement]. Your session can be total chaos, but...you can arrange it later and make it neat and tidy" (Churchman, 2025).

Another way to think about form is that it's the music's story, unfolding over time. But for the story to unfold we need to build a sense of connection among the sounds in the music, so that *each sound is relating to the others*. For example, we might set up the onset of one sound or effect such that it appears to trigger the onset of another. Such triggering to show connections creates a sense of inevitability and flow, making the music's form convey dynamism. Hopkins pursues this idea in all of his production work. "I'm always trying to create this feeling that you're hearing one sound because something else has happened" he says, "like there's an inextricable logic to this happening because it's like a story" (Churchman, 2025). Producer Beatrice Lewis (aka Beatrice) likewise conceptualizes form as a story that uses repetition:

> I kind of see [form] as a bit of a story with repeating phrases. And every time you tell the phrase again you're telling it in a slightly different way so it feels like you're getting to know the story deeper and deeper and deeper. I like to do that by repeating sections but when you repeat you change them slightly, or add in some new layers.
>
> *(Ableton, 2024a)*

The changes Beatrice brings to the repeating parts are subtle *parameter automations* on volume levels and effects. This helps parts dynamically evolve over time, making the music's form feel like a journey into an ever-deepening narrative:

> Another way I like to do that is...make different parameters on plug-ins, or using returns, so that repeated sounds feel like they're evolving. It's very subtle. But you can feel when all the sounds are changing and evolving and moving: it feels like they're alive. And I'm more wanting the music to sound like it's evolving. And so automating a lot of parameters on plug-ins, the track volumes, using a lot of [effect] returns...These are all going down to different delays, different reverbs that are EQ'd differently—all the kinds of different sounds that can make sounds sound like they're in different spaces. *How do I make an arrangement feel like it's telling a deeper and deeper and deeper story? How does it feel like a journey*?"
>
> *(Ableton, 2024b, italics added)*

Ways of developing ideas

From starting work on a track to finessing its details, a recurring question every producer answers through production workflows is, *How might I develop this?* Your answer depends on what you've begun with, and to some extent, where you want to go:

Did you begin with a beat, a series of chords, a single sample?
Are you aiming to make something beat-driven or beatless? Rhythmic or amorphous?

And how you develop what you've begun with steers you towards a musical destination:

Is the beat something you want to repeat or develop into something else?
Are these chords fine as they are, or should they be gradually modified?
Is a sample the track's focus, or will it become a source for a yet-to-be-created sound?

There are no right answers to such questions, because you don't know what will catch your attention until you're building things, trying out options. Building and trying options produces information that helps steer you towards the next step. Eventually, answering *How might I develop this?* leads you to the music's form. In the meantime, consider production strategies with a high probability of generating sounds and ideas.

Work with what's already at hand. As discussed in Chapter 4, it can be helpful to limit yourself to exploring the potentials of what you already have, whether it's a few chords, a beat, a sample, or a sound you like. If the beat lacks syncopation, chop it up, process it, and then process the processing. If the chords are uninspired, change their notes until they're original. If the sample is too obvious, mangle it into subtlety. If processing helped the beat, try it on the sample. The principle of working with what's already at hand is useful at every stage of production.

Loop the good bits. One way to identify a track's good bits is to loop a fragment of a beat, sequence, or sample to see if it reveals a magical moment. Such moments can be the basis of a track.[5] Good bits on loop inspire compelling larger forms.

Amplify textures. Sometimes a sound's beauty is hidden. Your job is to reveal it. A way to do this is to amplify overlooked aspects of sounds, such as their reverb tails. Amplify by increasing a track's volume, making surgical EQ boosts, or compressing the sound so that its quieter elements come to the foreground. Or combine these techniques.

Double the part. Copy a MIDI or audio part and process this doubled part into a new form. Turn a sampled voice into a pad, or re-pitch it into treble subliminal little sounds to fill out a mix. Any part can be doubled, and the double processed into something else.

Relate musical elements so they move together. As discussed in Chapter 2, a track's parts and parameters can modulate one another. Figure out modulations that make the music's elements move together in an interdependent call and response. This idea is well explained in a 2025 discussion on Reddit in which a producer named WizBiz92 suggests that we

> Gate things to other things, especially in a frequency specific way. Make them move together. The goal is to make one track's dynamics (or any other parameter) dependent on something tonal or harmonic in another track. Making things relate to each other in any way you can is a fantastic way to cohesively expand an idea.
>
> (*WizBiz92, 2025*)

Imagine how another producer might develop this. Ponder how another musician might expand your track. Are you being adventurous? Perhaps the problem is not that you lack material. Someone else might look at your screen and say, *But you have everything you need right here!*

Recursive production

No matter how elaborate a producer's musical system or what gear a producer uses, a powerful technique for creating sounds, parts, and structures is one that incorporates *recursion*. Recursion is the process of repeating items in a self-similar way. Self-similarity is abundant in the natural world, and recursive structures are found in things that are similar on different scales of resolution, like coastlines and trees (Figure 8.2).[6] In language, computer programming, and artistic practice, recursion is used as a method to solve problems by breaking them down into smaller instances of the same problem. As Paul Graham notes, recursion is one of two ways to create symmetry. (The other way is repetition.) "Recursive refers to an algorithm that refers to itself," he explains, using an analogy. "A policeman's algorithm for interrogating people is recursive: ask the person if they know about the crime, or if they know anyone who does, and if they do, interrogate them too" (Graham, 2004: 299).

Think of music production as always recursive insofar that *each track contains within itself answers to its questions*. Recursion is a powerful algorithm for composing because it's an approach to building upon—or, to use Graham's analogy, a way of

FIGURE 8.2 A tree's self-similar structure

interrogating—what one already has to generate new musical material. A recursive approach incorporates self-similar copies of your track into itself so that its sonic whole is woven from a set of related sounds. Now let's consider some ways to think about and incorporate a recursive approach into production.

Recursion mindset. The most important mindset shift is to understand that, wherever you might be in a project, *everything you already have is fodder for what is yet to be.*[7] As we've discussed, chords contain melodies and bass parts. But they can also be input for more abstract sound design/textural output. This means that everything you do *after* having played some chords can be, in some way, a self-similar repetition of them. Any aspect of a workflow benefits from self-similarity. For example, if you begin with beats, apply the same recursive approach by recycling parts of their rhythms to generate other percussive elements.

Solving production problems recursively. Temporary workflow quandaries, such as *What sound do I use?* or *How should I arrange this?* or *How do I blend these sounds?* can be solved recursively, rather than additively. We've learned that instead of getting lost in the possibilities of incorporating ever-new but *unrelated* sounds into your project, we can create by *altering what we already have.* Jon Hopkins describes this approach when he says "Maybe one of the most common [production] methods I use is to take something that's there [in a track] and process it into the next thing" (Tape Notes, 2021). Similarly, producer Austin Collins (aka Au5) will duplicate an existing sound into variations of itself. This process creates a family of related sounds:

> What I usually like to do is I'll take one main sound, and I'll duplicate it and tweak that and change it into slightly different sound. And then duplicate that, tweak it, and change it into a slightly different sound. And that way everything kind of has the fullness and it sounds like it's from the same palette. It's not just random, completely

> outlandish contrasting sounds. It's like [the sounds] all come from the same family. It's a lot easier to create a bunch of sounds with diversity from a single sound and have it sound good, than create a bunch of different sounds completely independently—built completely in different ways from one another—and then trying to make those sounds work together cohesively.
>
> *(Brett, 2021: 98–99)*

As Hopkins' and Collins' approaches show, blending sounds together is easier when those sounds are related. There are ways to ensure that this is the case. For example, if you've crafted a keys sound for a track, make a point of saving whatever processing you used as an effect chain. Try to duplicate parts of this chain to make new chains for other sounds elsewhere—maybe a distortion effect used to texturize the keys can be re-used on drums. This self-similar technique of duplicating effects illustrates how the problem of making a keys sound becomes a future solution for making a drum sound.[8] And as we learned in Chapter 7, the duplicate effect needn't be an exact repetition of its original. Change a few parameters and save the variation. It's through this practice that even one's effect chains continue to evolve as self-similar copies of their former selves.

Recursive ways forward. Adopting a recursive approach to music production diffuses two questions—*Do I have enough material?* and *What should I do next?*—that can bog down a workflow and induce anxiety. Connecting to Chapter 6's discussion of forward engineering, a recursive approach teaches the producer to forward engineer by turning materials upon themselves. This principle even applies to musical form. A track's form is recursive when it takes cues from the structural peculiarities of its component parts. What this means is that if I've begun a piece by improvising chords, the serendipities and errors in my playing are potentially as meaningful as the chords themselves. Perhaps my dynamics are inconsistent, my timing is loose, I awkwardly pause between phrases, and I repeat motives too many times. These quirks of performance can lead to interesting and non-obvious musical forms. "An ideal composition" notes producer Thomas Köner, "is ambiguous" (Fischer, 2022b). As an alternative to using conventional musical forms to anchor abstract sound design, we can find form by scaling up the music's component parts into a macro structure. As producer Felix Manuel (aka Djrum) reminds us regarding the fractal possibilities of production, by "reusing the same stuff over and over again…you can make so much out of the same material". New ideas, he says, "exist in the track already. It's endless because you put a small little [idea] in and you get so much out of it. One thing leads on to another" (MusicRadar Tech, 2025).

In sum, recursion is a way of looking inward rather than outward to solve questions about a track's sounds, parts, and structure by creating similarities on different scales of resolution. Elegant and easy to implement, recursion frames creativity in terms of recognizing the generative and self-similar properties of the artistic process itself. Sometimes a single sound really is enough to generate a whole piece.

Scaling

Once I have a mini-composition that feels like a possible track, I try to scale it up in some way. I'll layer other sounds to create timbral contrast, process what I have to change its texture, add a bass, maybe some kind of beat, use these MIDI parts to trigger those other sounds, and so on. Scaling up a track changes it, usually for the better, but one needs to consider what size is most appropriate for the track's material. Essayist John McPhee speaks of the effect of changes of scale on the impact of a work. "The idea" says McPhee, is that "a piece of writing ought not to be planned for a given size but developed to the length most suitable to the material, and no farther" (McPhee, 2022).

Scaling is a response to the question: *How can I transform this music over time?* It entails all the ways to build upon, extend, augment, and multiply an idea to realize a form suitable to one's material. Scaling techniques are many, including repetition, multiplication of parts, call and response, counterpoint, and effects processing. In the DAW these techniques can be realized simply enough through automated means—from copy and paste to transpose and invert, from sample and resample to mute, drag, and stretch. Scaling returns us to the question at the outset of this chapter about how to make a cohesive track. It's scaling that turns the 1-bar loop into the 32-bar phrase, a melody into a chord progression, a chord into an endless sequence, a one-shot sample into a drone horizon. The next time you sense an interesting musical idea-in-progress, try ways of scaling the idea to reveal its maximum expressiveness.

Scaling is a technique, but it's also a conceptual tool in that just *thinking about* scaling helps the producer hear even the smallest ideas as seeds for larger forms. For example, guitarist and producer Rachika Nayar thinks about scaling as a kind of prism for recombining her musical materials:

> That's how I think of mutating my guitar lines and sending it through these layers of recombination. It's like putting it through a prism; exploring a single idea through all these multiplying perspectives [and] seeing one thing in hundreds of its different facets and selves.
>
> *(Lockie, 2021)*

Seeing one thing in hundreds of its different facets brings us to another scaling approach, which is *doing the minimum to produce the maximum result*. Echoing the discussion of minimalisms and constraints in Chapter 4, take what you've recorded and declare, *from these notes, and only these notes, everything flows*. If you want another harmony or melody, it will come from here. Bass? From here. An entirely new sound or a large-scale structure? From here. This constraint brings to mind producer and film composer Cliff Martinez's *one thing* concept, which is about finding ways to derive a maximum of expression from a single sound:

> I usually started with the idea that I would score an entire scene with only one motif/sound/patch. When that didn't cut it, I would slow the motif by half, pitch it down an octave and add it to the piece as a bass element. Then I would try to squeeze as much expression out of the controllers and only when and if that still didn't complete the piece...I would add a new element.
>
> *(Heavyocity Media, n.d.)*

In sum, the *minimum into the maximum* concept is an elegant solution to scaling's open-ended problem. Its elegance comes from its self-similarity: there's beauty to deriving all of one's materials from an improvisation, or all of one's sounds from what's already at hand. When we work this way, we know that the parts with which we've augmented an initial gesture are *related* to it. As discussed above, this gives the music a fractal quality.

Prolation canons

Think about what keeps your attention when you encounter a music. Is it its

instrumentation and timbre world?
performers (human or machine)?
melodies?
chords and harmonies?
feel and vibe?
dimensionality?
low end?
beat?
Does the music slow you down or hype you up?
Does it invite you to concentrate on its processes, or does it percolate in the background?
Does it give you a reason to return to it, again and again?

Compelling musics create tensions, whose emotional magic doesn't diminish over time. We can think of these tensions as taut rubber bands in a state of stretch that contain energy. Musical tensions are enchanting because they keep changing and evolving "as the musical material is worked out" (Oram, 2016: 27), their dynamism arising through a series of mini feedback loops. Part steady-state and part ever-evolving, compelling musics dynamically maintain themselves like a runner in motion, manifesting a "sea of tension" (Crumley, 2020).

As seas of tension, compelling musics are constructed so that *parts relate to one another in measured ways*. One ancient counterpoint technique for achieving such tension that dates back to the 14th century is the *prolation* canon (or *mensuration* canon). Here a main melody is accompanied by imitations of it in other voices or parts. The tension created in musics built upon prolation canons comes from the fact that these imitations of the main melody happen *at different rates of speed* (or prolation). For example, an accompanying voice *extends* the note values of the main melody, through a process of *augmentation*, or *reduces* the melody's note values, through a process of *diminution*. Prolation counterpoint is one of music's most beguiling tensions and can be as useful in electronic music production as it was in Renaissance choral music. Try it yourself: write a melody, copy it, slow the copy to half speed, and listen to the counterpoint. It almost always sounds interesting.

The prolation technique evokes structural and design approaches in other realms. Think of an architect expressing a single idea in various facets of a building whose details proportionally relate to one another, a novelist weaving a broad storyline via several plots unfolding at different paces, or a chef preparing a single ingredient in multiple ways to construct a dish. In each of these scenarios the artist explores the fractal-like potentials of a set of materials, recognizing that even the smallest gesture, as we learned in Chapter 2, contains within itself a world of potential relations.

Transitions

"Transition is the very heart of inspiration and creativity" writes the inventor of the Rubik's Cube, Ernő Rubik in his autobiography (Rubik, 2020). Solving the Cube, a 3-dimensional object, is about recognizing patterns of color and then rotating the Cube's pieces to resolve those patterns into a single color per Cube side. The solving process is a transitioning of one pattern into another and then another, creating endless six-color combinations along the way. To solve the Cube is to unscramble it back to unity.

Recalling Chapter 1's discussion of beauty as symmetry, so production, too, is a kind of unscrambling of a track's constituent elements, devising transitions that take sounds on a journey, ushering the music from a here to a there. There are many forms of transition. A harmonic transition is one chord "moving" to another (we say moving, even though a chord progression is only metaphorical motion), or a cadence to end a phrase, such as V-I progression which transitions from a sense of tension-anticipation to a sense of stability-resolution. (This is why we say, metaphorically, that a I chord "resolves" the V.)

Transitions also connect one section of a track to another. A transition can take the form of a modulation or common tone, a drum fill, a drop or effects filter sweep, or the gradual addition or subtraction of parts. Transitions are such an integral part of music production that they now have their own plug-ins in the DAW. For example, in 2023 Baby Audio and production YouTuber Andrew Huang released Transit, a device meant to simplify the creation of transitions by combining multiple effects, whose parameters can be linked to a single macro control.

While transition plug-ins can achieve extraordinary sonic effects, a more fundamental kind of transition problem faced by the producer is: How do I develop a simple musical idea—a rhythm or a beguiling timbre—into something more robust? This is where we can learn from Rubik's Cube. One doesn't solve the puzzle simply by repeating patterns; transition requires *transforming* one pattern into another. In his book, Rubik evokes Christopher Alexander's alive-at-many-levels-at-once concept, looking to nature's fluidity and connectivity as an ecological model for thinking about design and creativity generally. "At its core, design is the link to nature for artificial objects" he says. "Nature does not know strict borders or barriers; it only knows transition" (Rubik, 2020: 56).

Producers often build musical structures upon mutually distinct, yet interconnected patterns layered one upon another: chords, melodies, rhythms, and timbres each doing different things at different speeds simultaneously to make a composite, interpenetrating *meta-pattern*. The composing that is producing—and the set of competencies we've come to understand as *omnimusical*—is fundamentally about figuring out ways of achieving such meta-patterns using one's musical system. Moreover, as producers learn how to use pattern-making tools, they undergo their own transition as students of the creative process. Producer Austin Collins (aka Au5) insightfully explains this metacognitive transformation. "It's not just learning the tool; it's learning how you interact with the tool—the relationship you have with synths and effects", he says (DAW Nation, 2020).

In music production there are as many kinds of transition as there are ways of manipulating sounds. Sometimes a musical process is its own transition: parts multiply as prolation canons, a track gradually slows down or erodes into noise, moves from mono to stereo, or melts into a drone. Other times transitions are finessed into the music as intricate automations, effect changes that suggest stories, or the recursive ear candy Kieran Hebden (aka Four Tet) in Chapter 5 called *little subliminal sounds*. Sometimes producers disguise the seams of their music so transition points are felt, but aren't obvious in how they work. Other times dramatic filter sweeps, drops, and fade-out transitions are a music's focal points. In sum, think of every track as a Cube of possibilities and your transitions like twists of the puzzle. Transitions are momentum processes that change the music from one state to another, like nature in its continuous flow.

Notes

1 Alexander elaborates on the scene: "It is the changing arrangement of the sky, the clouds, the flowers, leaves, the faces around us, the dazzling geometrical coherence, together with its meanings in our minds. But this geometry which means so much, which makes us feel the presence or order so clearly—we do not have a language for it" (Alexander, 2002: 9).

2 Some classic resources on the topic include Kubik (1962), Chernoff (1979), Hebdige (1987), Keil (1987), Collins (1996), Eshun (1999), and Arom (2004).

3 As Tim Rutherford-Johnson observed in 2013, producers including Kieran Hebden (aka Four Tet), Carsten Nicolai (aka Alva Noto), Richie Hawtin, and Nobukazu Takemura have acknowledged minimalism's influence. "Producers had begun to create a new form of techno that was more attuned to minute processes of variation and evolution" he writes (Rutherford-Johnson, 2013).

4 Ableton Live's MIDI transformation and generative tools implement this idea. See Ableton Live's manual (Ableton, 2024: 242–259).

5 I use this approach to make cycles versions of my tracks, turning fragments into new music.

6 Self-similar structures are also known as *fractals*. The term (from the Latin *frāctus* which means "fractured") was coined by mathematician Benoit Mandelbrot to describe "many of the irregular and fragmented patterns around us" in nature whose structures are "identical at all scales" (Mandelbrot, 1983: 1).

7 Composer David Cope, whose proto-AI Emmy software we discussed in the Interlude, says that "every work of music contains a set of instructions for creating different but highly related replications of itself" (Hochberg, 2014).

8 Engineers John Kuprenas and Matthew Frederick describe this process as adding tools to one's toolbox: "Add the tools you develop while solving each problem to your toolbox, to use on future problems. More important, add to your toolbox *the methods by which you discovered the new tools*" (Kuprenas and Frederick, 2018: 5, italics added).

9

ITERATING TOWARDS A QUALITY WITHOUT A NAME

Introduction

In a 2022 Ask Me Anything on Twitch, Sean Booth of Autechre explained an iterative process that underlies the group's music making. The process involves answering, as frankly as possible, a fundamental question about quality: *Is this shit or not?* Autechre make music by trying ideas, evaluating them, taking next steps, and repeating the process until they hear something that sounds good. Booth elaborates:

> What I tend to do is work with a very basic idea. Everything is drafts. It's drafts all the way down...You do something and you think, *Is this good? Yes or no? If no, what do I change to make it good?* You keep doing that and at some point you think, *Yes, okay that's good. So what does that then suggest that I put with it?* And then you might have what you call an idea. So you might think, *Well, I could try this or try that*, and at some point you might hit on something that works. So the tracks grow based on me saying, *Is this shit or not*, basically.
>
> *(Booth, 2022a)*

Notice the clarity of Booth's thinking about what's important in music production. Autechre work with a musical system built out of Max patches developed over decades of trial and error, a system that may seem a world apart from conventional conceptions of production. But theirs is actually a relatable method: work with simple ideas, ask yourself if they're good or not, and try ways of changing and developing these ideas through gradual refinement. Autechre's method depends, most of all, on trusting their taste (earned through years of experience) to figure out ways of moving their music forward. Producing music is work—it's drafts all the way down. The point is that even if Autechre isn't your kind of music, take inspiration from their humanistic aesthetics. It's not AI generating the duo's alien sound worlds. It's musicianship encountering technology, effort's results over time.

DOI: 10.4324/9781003585145-11

Iteration

Drafts all the way down is another way of talking about *iteration*, the production strategy of growing music by tinkering with the goal of moving a track ever closer to having the right form for its feeling and right feeling for its form. The word comes the Latin verb *iterum* or "again" and so to iterate is to repeat a process with the goal of refining something or bringing it towards a desired result. Iteration is everywhere we make things—from practicing a musical instrument to editing prose—because it's the most reliable way of understanding what we're trying to accomplish. Six examples from a thousand possibilities: begin with a basic drum sound and complexify it, give a flat keyboard depth, turn a dry texture into lush ambience, play with perspective, experiment with rhythmic energies, reduce and refine a musical line until it shines. Return to the music again and again to sharpen the elements already present, nudging them towards interestingness. Iteration trusts its process: polish the demo, finesse the first take, keep versioning the version. And iteration is unflappably optimistic, believing that more is possible when you practice repetition without repeating yourself, that you can use what you notice to shape changes in the work, and that quality compounds over time.

Iterating involves trial-and-error experimenting, and specifically, *tinkering*, which means trying to improve something in a casual, playful way. Picture a sculptor chipping away at a form, an artist sketching, a cook varying a recipe, or a writer moving sentences around until they sit right. (In medieval times, a tinkerer was a craftsman who worked with tin, using the cheap metal to mend pots and pans.) In his book *Antifragile*, Nassim Taleb defines tinkering as the process of making a series of small, low-risk practical experiments. As we learned in Chapter 6, tinkering is the most empirical way to learn about music production because it provides instant feedback in the form of small losses, as well as the occasional insight. "While engaged in tinkering" Taleb says, "you incur a lot of small losses, then once in a while you find something rather significant" (Taleb, 2012: 236). Experimental tinkering plays with different versions or formulations of an object or idea. In his book about developing the iPhone, software engineer Ken Kocienda describes the *demo*, the tinkering process Apple engineers use to improve products. Teams built software and hardware demo prototypes, demo-ed them to one another, and shared critiques to accelerate their learning about what did and didn't work. "We improved our demos in incremental steps" explains Kocienda. "We evolved our work by slowly converging on better versions of the vision" (Kocienda, 2018: 221).

Examples abound of how music producers are, first and foremost, experimental tinkerers who work in incremental steps. Producer and guitarist Julia Reidy speaks of her workflow's phases that "involve experimenting with the material, allowing myself to have long thought processes and taking lots of little detours and tangents" (Udvardyova, 2021). Producer Lawrence English speaks of how experimentation leads him to "come across elements or sounds that spark entire chains of events" (English, n.d.). Producer Chris Clark (aka Clark) sums up his iterative tinkering as "endless processing in Ableton. Just freezing, printing, freezing that, printing that, re-pitching, transcribing that into sample libraries, bouncing that" (Fallon, 2014). Producer A.G. Cook likes to repeat a single musical idea, day after day, iterating to get closer to the idea's ideal version and take pressure off the process. His goal is to not get too attached to results, until one day the idea coalesces: "I would rather do many of the same attempt at something and delete some of them, forget others, and then on one random day, nail it and make the perfect version of that" he says (Tape Notes, 2024a). Similar to A.G. Cook, producer Felix Manuel (aka Djrum)'s iterative process involves making multiple versions of a track over many sessions, each version an opportunity to tinker with new ideas. The track evolves as Manuel preserves the best attributes of each successive version:

> My sessions in Ableton end up with a version of the tune and then I'll always copy and paste that, and do another version with a new combination. I'll keep each on, trying new things, often taking the best bits from each version. If you hear a sound that comes into the tune once and never again, that might be a really important part of another version that I didn't want to use a lot but didn't want to lose it altogether. The end result is very fine-tuned and heavily produced over lots of sessions. In a way, when you produce programmed music, you're not capturing a moment.
>
> *(XLR8R, 2018)*

Producers' creative options expand in proportion to their practices of iteration. Put simply, we don't know what's possible until we tinker with lots of options to hear how they sound. As Manuel says, we're not capturing a moment, we're building one. This takes time, but results—what Taleb called *finding something rather significant*—are guaranteed. To return to Autechre, Booth explains that this is how one makes good work—by randomly trying out things until you notice something of quality:

> The thing I like about Max is that it lets me try out lots and lots of different things and then figure out which ones work. I think that's quite a productive way of working. A lot of people think that you have to have this strong conceptual idea that you then realize and that makes you creative. But I don't think being creative should be the goal, actually. I think making good work should be the goal. Quite often just doing a lot of random shit and then noticing that something's amazing can produce better work. That's what experience has shown.
>
> *(Booth, 2022)*

One way to explore iteration on a small scale is to change a single sound repeatedly until it becomes a compelling presence you never would have imagined had you not created it. With iteration, one's sounds have no limit. Reflecting on his workflows, producer Jon Hopkins observes a fact of sound designing, which is "that it almost doesn't matter what the source is, if you're using this amount of processing" (Tape Notes, 2021). Processing sound elasticizes its timbre; with some serendipity and skill a producer can turn *this* sound into almost any *other* sound. Furthermore, *new ideas typically emerge as we're tinkering* and noticing certain options that are close enough at hand to warrant trying, such as processing what we already have into something new. The producer builds up and strips down a track through improvised processing-into-the-next-thing moves that may never be replicated. With each iteration more possibilities reveal themselves, cued by moves already tried. Each time around an iteration's cycle it's as if an adjacent possible is whispering, *And have you thought about this?* [1]

Even if we never repeat the same production experiments, over time iteration begins to imprint itself on us. We develop the equivalent of a muscle memory for knowing how to activate creative cascades within our musical system. We remember how we used reverb to create a pitched-down, oceanic space around a sound. We remember having made enough versions of this space to understand it as an effect useful in certain musical contexts. Iteration also led us to variations on the effect—how to widen a space, or make it more textured, and so on. Our production iterations are like branches of an idea tree. "Knowledge grows fractally" (Graham, 2023a) in that new discovery branches continue growing off of the tree. Thus, iteration amplifies an important general compositional concept, which is that starting points matter, but only to a point. A track's beginning might have been a quick improvisation or a carefully worked-out rhythm. It doesn't matter what the music's source is because once you have a starting point—what Booth calls a *Yes, okay that's good* realization—iteration will lead you to new discovery branches.

Now let's return one last time to Christopher Alexander's alive-at-many-levels-at-once idea, vividly illustrated in Chapter 2's beach scene. This idea appears throughout this book because it encapsulates the force and mystery of electronic music production. But there's more to the idea. Alive at many levels at once generates what Alexander called a quality without a name, which refers to the dynamics of a living system of interdependent patterns that are stable, yet varied. "The quality without a name occurs" he says, "not when an isolated pattern occurs, but when an entire system of patterns, interdependent, at many levels, is all stable and alive (Alexander, 1979: 131). A quality without a name is an apt description of music production's aesthetic of energy. The producers we aspire to emulate and learn from pursue this quality. As they shape sounds and build stable yet varied music, they aim to conjure the sense of interconnectedness, vividness, and vibrancy that comes from sounds alive at many levels at once. In sum, the way to perpetually evolve one's workflow and skills in pursuit of a quality without a name is to keep iterating. That's it. There's no magic hack, no AI solution—music production is work. Creativity happens through iteration, composed through small changes effected over time. Serum synthesizer designer Steve Duda sums up the idea. "You're sort of incrementally just modifying" he says, "into existence, into something that's yours" (The Mr. Bill Podcast, 2020).

The value of making a version 1.0

The concept of creating and releasing a version 1.0 originated in computer programming circles, where sharing software-in-progress helps one learn how it works and identify errors that need fixing. In his book *Hackers and Painters*, Paul Graham explains that sharing a version 1.0 is the best way to get feedback. He urges programmers and artists alike to "get a version 1.0 out there as soon as you can. Until you have some users to measure, you're optimizing based on guesses" (Graham, 2004: 127).

Following Graham, provisionally finish each track idea as a workable version, mix it, and bounce it to audio. This is your version 1.0, a tangible demo you can listen to, assess, think about, and iterate upon. You're not releasing version 1.0 of a track to the public necessarily, but building a 1.0 for yourself to get a sense of what it is and where you might take it. As you iterate this process of quickly getting tracks into shape you'll become better at knowing what they need. Instead trying a thousand ideas, you'll learn to pursue the first close at hand idea that's promising. If the music suggests an intriguing mood or direction, you can move onto the next iteration, whatever that is.

On an Ableton sub-Reddit, in response to another producer's question about the best way to spend an hour a day practicing production, a user (nadalska) offers a succinct, iteration-based answer: "Give yourself some limitations, like only using 4 tracks or a particular set of sounds and just make a track as quick as possible. Don't have any expectations, repeat" (darmud, 2024). Thinking in terms of version 1.0 and then practicing doing so—by having limitations and working quickly without expectations—brings momentum to our producing. Building version 1.0s over and over sharpens our iterating. Now we're practicing Bernstein's *repetition without repetition* concept, varying our workflows each time we make music. Our workflows evolve, just like us.

Conclusion

Summing up, this book has explored what I consider to be the principles of electronic music production, illustrated with ideas and techniques of numerous musicians working across a range of styles. My hope is that in the book's various perspectives readers find ideas to think about and incorporate into their own production work. And so, to recap the main themes, I urge you to

> embrace serendipities, errors, and happy accidents; create a living musical system alive at many levels at once; play the DAW like an instrument; tweak sounds, modulate your modulators; trust the musical gesture; devise minimalisms and constraints by using sounds and instruments closest at hand; commit to what's working right now; incorporate your sense of touch; improvise constantly; bloom other parts from chords; drum non-obvious rhythms; get each layer right; embrace the synthetic but evoke the acoustic (or vice versa); contrast perspectives and textures (i.e. treble v bass, dry v resonant, mono v stereo, foreground v background, clean v gnarly, freeform v measured); invent production prompts; attune to sound feels; resample what's already there; amplify artifacts; make field recordings and drones; connect effects into effect chains to invent new affects; develop dynamic structures by elegant (simple) methods; find the perfect loop; get off the quantized grid; create tonal evolutions; think of musical form as a story; scale up recursively; unfold parts at different rates of speed; develop multiple versions of the same idea; create and refine through iteration; suspend self-judgment; seek a quality without a name; and always, always save your work.

Note

1 Steven Johnson explains Stuart Kauffman's *adjacent possible* concept: "At any given moment in evolution—of life, of natural systems, or of cultural systems—there's a space of possibility that surrounds any current configuration of things" (Kelly and Johnson, 2018).

REFERENCES

Ableton. (2024). Ableton Reference Manual Version 12. *Ableton.* Available from: <https://www.ableton.com/en/manual/welcome-to-live/> [Accessed 12 February, 2025]

Ableton. (2024a). Find out what's possible in Live 12. *Ableton.* Available from: <https://www.ableton.com/en/live/ > [Accessed 12 February, 2025]

Ableton. (2024b). Made in Ableton Live: Beatrice on creative vocal sampling techniques and the power of limitations. *Ableton.* Available from: <https://www.youtube.com/watch?v=7RpeHJFAZrU&t=1722s> [Accessed 28 April, 2025]

Ableton. (2017). One Thing: Ben Lukas Boysen - Dynamic Layering. *Ableton.* Available from: <https://www.youtube.com/watch?v=eT1Ouhd-p1Q> [Accessed 28 April, 2025]

Ableton. (2017a). *Bonobo: The Path to Migration. Ableton.* Available from: <https://www.ableton.com/en/blog/bonobo-path-to-migration/> [Accessed 28 April, 2025]

Adams, T. (2010). David Cope: 'You Pushed the Button and out Came Hundreds and Thousands of Sonatas .' *The Guardian.* Available from: <https://www.theguardian.com/technology/2010/jul/11/david-cope-computer-composer> [Accessed 28 April, 2025]

Abram, D. (2010). *Becoming Animal: An Earthly Cosmology.* New York, Vintage.

Adrià, F. (2010). *A Day at elBulli.* London, Phaidon.

A-Ha. (2011). A-Ha - Take on Me - Paulstretch 800% Slower Version. *YouTube.* Available from: <https://www.youtube.com/watch?v=ZmD7z7uH-Hg> [Accessed 28 April, 2025].

AIR podcast. (2024). JakoJako, Episode 68. Available from: <https://soundcloud.com/airpodcast/air-068-jakojako> [Accessed 28 April, 2025]

Alexander, C. (2002). *The Nature of Order: An Essay on the Art of Building and the Nature of the Universe*, Book 1 – *The Phenomenon of Life.* Oxford, Oxford University Press.

Alexander, C. (1979). *The Timeless Way of Building.* Oxford, Oxford University Press.

Allen, J. (2021). Celer: Creating the Feeling of Place, and Questioning Where That Is. *Cyclic Defrost.* Available from: <https://www.cyclicdefrost.com/2021/07/celer-creating-the-feeling-of-place-and-questioning-where-that-is/> [Accessed 28 April, 2025]

Anagnostellis, A. (2016). Mr. Carmack on the Power of Technology in Music Production… and Socks. *The Music.* Available from: <https://themusic.com.au/features/interview-mr-carmack/X697c3J1dHc/07-01-16> [Accessed 28 April, 2025]

Armstrong, R. (1971). *The Affecting Presence: An Essay in Humanistic Anthropology.* Urbana-Champaign, University of Illinois Press.

Arnott, B. (2013). Huerco S Interview: Conducting Experiments & Discovering Artefacts. *Juno*. Available from: <https://www.juno.co.uk/junodaily/2013/01/07/conducting-experiments-discovering-artefacts-with-huerco-s/> [Accessed 28 April, 2025]

Arom, Simha. (2004). *African Polyphony and Polyrhythm: Musical Structure and Methodology*. Cambridge, Cambridge University Press.

Asparouhova, N. (2024). Does meditation experience improve success with the jhanas? Nadia Asparouhova. Available from: <https://nadia.xyz/meditation-experience> [Accessed 28 April, 2025]

Assar, V. (2012). Squarepusher: is Tom Jenkinson. *Tape Op*. Available from: <https://tapeop.com/interviews/89/squarepusher/> [Accessed 28 April, 2025].

Bach, G. (2003). The Extra-Digital Axis Mundi: Myth, Magic and Metaphor in Laptop Music. *Contemporary Music Review*, vol. 22, no. 4: pp. 3–9.

Baker, J.A. (1967). *The Peregrine*. New York, William Collins.

Bartos, K. (2022). *The Sound of the Machine*. London, Omnibus Press.

Bates, L. (2022). The Process: The Caretaker. *The Believer*. Available from: <https://www.thebeliever.net/logger/the-process-the-caretaker/> [Accessed 28 April, 2025]

Bateson, G. (2000). *Steps to an Ecology of Mind*. Chicago, University of Chicago Press.

Beaumont-Thomas, B. (2023). We Soon Won't Tell the Difference between AI and Human Music – So Can Pop Survive? *The Guardian*. Available from: <https://www.theguardian.com/music/2023/apr/19/ai-human-music-pop-drake-kanye-west-the-weeknd> [Accessed 28 April, 2025]

Bell Telephone Laboratories, Inc. (1960). Music from Mathematics. Vinyl LP.

Berleant, A. (1999). Notes for a Phenomenology of Musical Performance. *Philosophy of Music Education Review*, Vol. 7, No. 2: pp. 73–79.

Bernstein, N. (1967). *The co-ordination and regulation of movements*. Oxford, Pergamon Press.

Blitz, M. (2016). How Sam Phillips Invented the Sound of Rock and Roll. *Popular Mechanics*. Available from: <https://www.popularmechanics.com/culture/music/a22237/sam-phillips-sun-studio/> [Accessed 28 April, 2025]

Bonnet, F. (2020). *The Music to Come*. Britany, Shelter Press.

Bonnet, F. (2020a). Sensitive States of Perception - Kali Malone interviewed by François Bonnet. *Sonic Acts*. Available from: <https://sonicacts.com/archive/sensititve-states-of-perception-with-kali-malone> [Accessed 28 April, 2025]

Booth, S. (2022). Autechre 4utechre Sean QUICK AMA Twitch 6/5/2022 (Part 1). *YouTube*. Available from: <https://www.youtube.com/watch?v=rtyY_3p7vkE&t=1606s> [Accessed 28 April, 2025]

Booth, S. (2022a). Autechre 4utechre Sean QUICK AMA Twitch 6/5/2022 (Part 2). *YouTube*. Available from: <https://www.youtube.com/watch?v=OdZv1PNxDcs> [Accessed 28 April, 2025]

Booth, S. (2022b). Autechre 4utechre Sean QUICK AMA Twitch 6/5/2022 (Part 4). *YouTube*. Available from: <https://www.youtube.com/watch?v=o3u5ESOW5Cw> [Accessed 28 April, 2025]

Brand, S. (1995). *How Buildings Learn*. New York, Penguin Books.

Brett, T. (2021). *The Creative Electronic Music Producer*. London, Routledge.

Brett, T. (2025). database (a cache of perceptions). brettworks. Available from: https://brettworks.com/music-production-database/ [Accessed 4 May, 2025]

Bridle, J. (2022). *Ways of Being: Animals, Plants, Machines: The Search for a Planetary Intelligence*. New York, Farrar, Straus and Giroux.

Broomfield, D. (2023). Lee Gamble: Models. *Pitchfork*. Available from: <https://pitchfork.com/reviews/albums/lee-gamble-models/> [Accessed 28 April, 2025]

Bruce-Jones, H. (2018). Listen to Actress's Mini-Album with A.I. Sprite Young Paint. *Fact Magazine*. Available from: <https://www.factmag.com/2018/09/28/listen-actress-young-paint-mini-album/> [Accessed 28 April, 2025]

Bruce-Jones, H. (2023). Interview: Malibu. *Fact Magazine*. Available from: <https://www.factmag.com/2023/11/03/malibu-interview/> [Accessed 28 April, 2025]

Buchanan, J. (2024). Logic Pro: Ideas Machine Template. *Jono Buchanan Music*. Available from: <https://www.youtube.com/watch?v=GHAHVW80PMo> [Accessed 28 April, 2025]

Burrows, D. (1997). "A Dynamical Systems Perspective on Music." *The Journal of Musicology*, Vol. 15, No. 4, pp. 529–545. Available from: <doi:10.2307/764006" \t "_blank"doi:10.2307/764006>

Buskin, R. (1995). Anne Dudley: Art of Noise. *Sound On Sound*. Available from: <https://www.soundonsound.com/people/ann-dudley-art-noise?amp> [Accessed 28 April, 2025]

Cableguys. (2025). Snapback. *Cableguys*. Available from: <https://www.cableguys.com/snapback> [Accessed 28 April, 2025]

Cain, R. (2020). 'Looking For Details in Sound and Researching My Surroundings': Talking Tech with Upsammy. *Stamp The Wax*. Available from: <https://www.stampthewax.com/2020/06/17/talking-tech-upsammy/> [Accessed 28 April, 2025]

Calvino, I. (1985). *Mr. Palomar*. New York, Harcourt Brace Jovanovich, Inc.

Canby, E. (1955). Synthesized Music, by Edward Tatnall Canby. *Harper's Magazine*. Available from: <https://harpers.org/archive/1955/09/synthesized-music/> [Accessed 28 April, 2025]

Cardew, B. (2020). Solid Gold: How the Beat Science on Photek's 'Modus Operandi' Redefined Drum & Bass. *DJ*. Available from: <https://djmag.com/longreads/solid-gold-how-beat-science-photeks-modus-operandi-redefined-drum-bass> [Accessed 28 April, 2025]

Carr, N. (2025). *Superbloom: How Technologies of Connection Tear Us Apart*. New York, W.W. Norton & Company.

Cascone, K. (2002). The Aesthetics of Failure. *Ccapitalia.net*. Available from: <https://www.ccapitalia.net/reso/articulos/cascone/aesthetics_failure.htm> [Accessed 28 April, 2025]

Catmull, E. (2014). *Creativity, Inc*. New York, Random House.

Cave, N. (2023). Chat GPT What Do You Think? *The Red Hand Files*. Available from: <https://www.theredhandfiles.com/chat-gpt-what-do-you-think/> [Accessed 28 April, 2025]

Charnas, D. (2022). *Dilla Time: The Life and Afterlife of J Dilla, the Hip-Hop Producer Who Reinvented Rhythm*. New York, MCD.

Chaudhuri, A. (2021). *Finding the Raga: An Improvisation on Indian Music*. New York Review Books.

Chernoff, John Miller. (1979). *African Rhythm and African Sensibility*. Chicago, University of Chicago Press.

Chiang, T. (2023). ChatGPT Is a Blurry JPEG of the Web. *The New Yorker*. Available from: <https://www.newyorker.com/tech/annals-of-technology/chatgpt-is-a-blurry-jpeg-of-the-web> [Accessed 28 April, 2025]

Chiang, T. (2024). Why A.I. Isn't Going to Make Art. *The New Yorker*. Available from: <https://www.newyorker.com/culture/the-weekend-essay/why-ai-isnt-going-to-make-art> [Accessed 28 April, 2025]

Churchman, H. (2022). Tennyson: Storytelling with Production. *Ableton*. Available from: <https://www.ableton.com/en/blog/tennyson-storytelling-with-production/> [Accessed 28 April, 2025]

Churchman, H. (2025). Jon Hopkins: No Comfort Without Discomfort. *Ableton*. Available from: <https://www.ableton.com/en/blog/jon-hopkins/> [Accessed 28 April, 2025]

Clifton, T. (1983). *Music as heard: a study in applied phenomenology*. New Haven, Yale University Press.

Cohen, L. (1992). Anthem. Available from: <https://genius.com/Leonard-cohen-anthem-lyrics> [Accessed 28 April, 2025]

Coldeway, D. (2022). Try 'Riffusion,' an AI model that composes music by visualizing it. *TechCrunch*. Available from: <https://techcrunch.com/2022/12/15/try-riffusion-an-ai-model-that-composes-music-by-visualizing-it/> [Accessed 28 April, 2025]

Collins, John. (1996). *Listening to the Silence: African Cross-Rhythms*. Films Media Group. DVD.

Computer Music. (2017). Dom & Roland - Producer Masterclass with Computer Music (Part 1). *Computer Music*. Available from: <https://youtu.be/P8wiyllawYI> [Accessed 28 April, 2025]

Computer Music. (2013). Rusko - Producer Masterclass." *Computer Music*. Available from: <https://www.youtube.com/watch?v=ZCfn8P2kBXM> [Accessed 28 April, 2025]

Computer Music, (2012). Interface DnB Producer Masterclass - Computer Music Magazine." *Computer Music*. Available from: <https://www.youtube.com/watch?v=PFpfNGB7N9k> [Accessed 28 April, 2025]

Coscarelli, J. 2023. "An A.I. Hit of Fake 'Drake' and 'The Weeknd' Rattles the Music World." *The New York Times*, 19 (https://www.nytimes.com/2023/04/19/arts/music/ai-drake-the-weeknd-fake.html).

Cox, C. And C.Warner, eds. (2004). *Audio Culture: Readings In Modern Music*. New York, Continuum.

Cragg, M. (2022). *In-Demand Producer Fred Again.: 'I Was Fortunate Not to Be Good at Anything Else.' The Guardian*. Available from: <https://www.theguardian.com/music/2022/oct/17/fred-again-interview> [Accessed 28 April, 2025]

Crumley, E. (2020). Shane Benzie on the art of running. *Athletics Weekly*. Available from: <https://athleticsweekly.com/performance/shane-benzie-on-the-art-of-running-1039937770/> [Accessed 28 April, 2025]

Cunningham, M. (1994). "From Jams to James: Brian Eno." *The Mix*. Available from: <https://www.muzines.co.uk/articles/from-jams-to-james/8173>

Currin, G.H. (2021). Jon Hopkins's Psychedelic Journey to a New Way of Creating Music. *The New York Times*. Available from: <https://www.nytimes.com/2021/11/10/arts/music/jon-hopkins-music-for-psychedelic-therapy.html> [Accessed 28 April, 2025]

Danz, (2023). Interview with Tame Impala. *Synth History*. Available from: <https://www.synthhistory.com/post/interview-with-tame-impala> [Accessed 28 April, 2025]

darmud, (2024). *What's an Effective Way to Spend Time on Ableton for at Least an Hour Everyday for a Beginner Producer? Reddit*. Available from: <www.reddit.com/r/ableton/comments/1e0ty5f/whats_an_effective_way_to_spend_time_on_ableton/> [Accessed 28 April, 2025]

Davis. T. (2024). Upsammy: Crossing Strange Meridians. *Ableton*. Available from: <https://www.ableton.com/en/blog/upsammy-crossing-strange-meridians/?mtm_campaign=related-recommendation&mtm_source=blog&mtm_medium=related> [Accessed 28 April, 2025]

DAW Nation. (2020). Making Emotional Bass Music. *DAW Nation*. Available from: <https://www.youtube.com/watch?v=aalmnBvznRU> [Accessed 28 April, 2025]

Demers, J. (2010). *Listening through the Noise*. New York, Oxford University Press.

Denyer, R. (1986). No Presets Allowed. *Sound On Sound*. Available from: <https://www.muzines.co.uk/articles/no-presets-allowed/1639> [Accessed 28 April, 2025]

D'Errico, M. (2022). *Push: Software Design and the Cultural Politics of Music Production*. London, Oxford University Press.

Deshoqub. (2018). Autechre's Tsugi Interview Translated. *Reddit*. Available from: <www.reddit.com/r/autechre/comments/9of3dv/autechres_tsugi_interview_translated/> [Accessed 28 April, 2025]

dokev. (2023). Lanark Artefax how to and general technical pointers? *Elekronauts*. Available from: <https://www.elektronauts.com/t/lanark-artefax-how-to-and-general-technical-pointers/205930> [Accessed 28 April, 2025]

Doran, J. (2008). Revenge of the Synth: Benge Interviewed & Synth Porn Photo Gallery. *The Quietus*. Available from: <https://thequietus.com/interviews/benge-revenge-of-the-synth/> [Accessed 28 April, 2025]

Douglas, A. (2021). 'We Do It Because We Want to, Not Because We See Commercial Opportunities' - Urs Heckmann. *Attack Magazine*. Available from: <https://www.attackmagazine.com/features/interview/we-do-it-because-we-want-to-not-because-we-see-commercial-opportunities-urs-heckmann/#google_vignette> [Accessed 28 April, 2025]

Doyle, T. (2021). Fred Gibson Aka Fred Again. *Sound On Sound*. Available from: <https://www.soundonsound.com/people/fred-gibson-aka-fred-again> [Accessed 28 April, 2025]

Doyle, T. (2015). *Classic Tracks: A Guy Called Gerald 'Voodoo Ray.' Sound On Sound*. Available from: <https://www.soundonsound.com/techniques/classic-tracks-guy-called-gerald-voodoo-ray> [Accessed 28 April, 2025]

D'Silva, S. (2023). Cristobal Tapia de Veer: Hollywood Composer & Producer. *Sound On Sound*. Available from: <https://www.soundonsound.com/people/cristobal-tapia-de-veer> [Accessed 28 April, 2025]

Dunworth, L. (2022). *Brian Eno on Fred Again's 'Non-Linear' Musical Style: 'I Didn't Really Understand a Lot of What He Was Doing.' MusicTech*. Available from: <https://musictech.com/news/music/brian-eno-describes-fred-agains-unusual-songwriting-style/> [Accessed 28 April, 2025]

du Sautoy, M. (2019). What Are We Saying When We Call an Algorithm 'Creative'? *Literary Hub*. Available from: <https://lithub.com/what-are-we-saying-when-we-call-an-algorithm-creative/> [Accessed 28 April, 2025]

Elkind, R. (1972). *Liner notes for Sonic Seasonings*. Columbia Records PG 31234, compact disc.

Em, C. (2020). *Field Recording Workshop w/KMRU UNDER THE BRIDGE' ADE 2020*. Available from: <https://www.youtube.com/watch?v=B4F5PTbki_A> [Accessed 28 April, 2025]

English, L. (n.d.) *Fifteen Questions Interview with Lawrence English*. Available from: <https://15questions.net/interview/fifteen-questions-interview-lawrence-english/page-1/> [Accessed 28 April, 2025]

Eno, B. and B. Adriaanse. (2024). *What Art Does: An Unfinished Theory*. New York, Faber & Faber.

Eno. B. (1978). Ambient 1: Music For Airports. EG Records.

Ensoniq. (1992). Ensoniq ASR-10 Manual. Available from: <https://www.deepsonic.ch/deep/docs_manuals/ensoniq_asr10_manual.pdf> [Accessed 28 April, 2025]

Eshun, K. (1999). *More Brilliant Than the Sun*. London, Quartet Books.

Expressive E. (2025). Expressive MPE Polyphonic Synthesizer. Available from: <https://www.expressivee.com/2-osmose> [Accessed 28 April, 2025]

Faber, T. (2021). Decolonizing Electronic Music Starts with Its Software. *Pitchfork*. Available from: <https://pitchfork.com/thepitch/decolonizing-electronic-music-starts-with-its-software/> [Accessed 28 April, 2025]

Fallon, P. (2014). In the Studio: Clark. *XLR8R*. Available from: <https://xlr8r.com/gear/in-the-studio-clark/> [Accessed 28 April, 2025]

Feld, S. (2004). The Time of Bells, 1. VoxLox. CD.

Feld, S. (1991). Voices of the Rainforest: A Day in the Life of the Kaluli People. Rykodisc: B0000009NX. CD.

Fell, M. (2022). *Structure and Synthesis: The Anatomy of Practice*, Cambridge: MIT Press.

Ferriss, T. (2022). The Tim Ferriss Show, Episode 562. Available from: <https://podcasts.apple.com/us/ podcast/the-tim-ferriss-show/id863897795?i=1000547208870> [Accessed 28 April, 2025]

Finlayson, A. (2022). Friends, not Samples: The Soft Pink Truth on Collaboration and Pleasure. *Ableton*. Available from: <https://www.ableton.com/en/blog/friends-not-samples-the-soft-pink-truth-on-collaboration-and-pleasure/> [Accessed 28 April, 2025]

Fintoni, L. (2020). *Bedroom Beats & B-Sides: Instrumental Hip-Hop & Electronic Music at the Turn of the Century*. London, Velocity Press.

Fischer, T. (2023). Thomas Fehlmann: 'An Important Part of the Creative Process Is Forgetting.' *15 Questions*. Available from: <https://www.15questions.net/interview/thomas-fehlmann-about-production-technology-and-creativity/page-1/> [Accessed February 17, 2025]

Fischer, T. (2023a). JackoJacko about Production, Technology, and Creativity. *15 Questions*. Available from: <https://15questions.net/interview/jakojako-about-production-technology-and-creativity/page-1/> [Accessed 28 April, 2025]

Fischer, T. (2022). Nils Frahm About Space, Manipulation, and the Heart of the Artichoke. *15 Questions*. Available from: <https://www.15questions.net/interview/nils-frahm-about-space-manipulation-and-heart-artichoke/page-1/> [Accessed 28 April, 2025]

Fischer, T. (2022a). Fifteen Questions with Thomas Köner. *15 Questions*. Available from: <https://15questions.net/interview/fifteen-questions-thomas-koner/page-1/> [Accessed 28 April, 2025]

Fischer, T. (2022a). Patricia Wolf about a Synthesizer with Effortless Flow and Almost Limitless Possibilities. *15 Questions*. Available from: <https://15questions.net/interview/tools-creation-novation-summit/page-1/> [Accessed 28 April, 2025]

Fischer, T. (2022b). Fifteen Questions with Thomas Köner. *15 Questions*. Available from: <https://15questions.net/interview/fifteen-questions-thomas-koner/page-1/> [Accessed 28 April, 2025]

Fischer, T. (2021). Fifteen Questions Interview with Astrid Sonne: Transferring Energies. *15 Questions*. Available from: <https://www.15questions.net/interview/fifteen-questions-astrid-sonne/page-1/>. [Accessed March 11, 2025]

Fischer, T. (2012). Interview with Chichei Hatakeyama. *tofaki*. Available from: <http://www.tokafi.com/15questions/interview-chihei-hatakeyama/> [Accessed 28 April, 2025]

Fischer, T. (2012a). Interview with Efdemin. *tofaki*. Available from: <http://www.tokafi.com/15questions/interview-efdemin/> [Accessed 28 April, 2025]

Fischer, T. (n.d.). Fifteen Questions with Richard Skelton. *15 Questions*. Available from: <https://www.15questions.net/interview/fifteen-questions-interview-richard-skelton/page-3/> [Accessed 28 April, 2025]

Fisher, M. (2013). The Metaphysics of Crackle: Afrofuturism and Hauntology. *Dancecult: Journal of Electronic Dance Music Culture*, Vol. 5, no. 2: 42–55. Available from: <https://dj.dancecult.net/index.php/dancecult/article/view/378/391> [Accessed 28 April, 2025]

Fisher, M. (2012). Burial: Unedited Transcript. *The Wire*. Available from: <https://www.thewire.co.uk/in-writing/interviews/burial_unedited-transcript> [Accessed 28 April, 2025]

Fitzpatrick, R. (2023). 'Like brushing my teeth': how Michiru Aoyama writes, records and releases an album every day. *The Guardian*. Available from: <https://amp.theguardian.com/music/2023/dec/05/michiru-aoyama-ambient-album-every-day-interview> [Accessed 28 April, 2025]

Flint, T. (2008). Jean-Michel Jarre: 30 Years of Oxygène. *Sound On Sound*. Available from: <https://www.soundonsound.com/people/jean-michel-jarre-30-years-oxygene> [Accessed 28 April, 2025]

Fosl, L. (2018). Jan Jelinek. *Tiny Mix Tapes*. Available from: <https://www.tinymixtapes.com/features/jan-jelinek-zwischen> [Accessed 28 April, 2025]

Friedlander, E. (2019). How Holly Herndon and Her AI Baby Spawned a New Kind of Folk Music. *The Fader*. Available from: <https://www.thefader.com/2019/05/21/holly-herndon-proto-ai-spawn-interview> [Accessed 28 April, 2025]

Future Music. (2022). 'It Was Just a Pure Outpouring of Music': Orbital on How They Made In Sides. *MusicRadar*. Available from: <https://www.musicradar.com/news/classic-album-orbital-in-sides> [Accessed 28 April, 2025]

Future Music. (2022a). Classic Interview: Oneohtrix Point Never. *Future Music*. Available from: <https://www.musicradar.com/news/oneohtrix-point-never-drums-synths> [Accessed 28 April, 2025]

Future Music. (2020). Caribou: 'Even If a Hardware Sampler Has a Slightly Different Sound, It's Too Much of a Faff .' *MusicRadar*. Available from: <https://www.musicradar.com/news/caribou-even-if-a-hardware-sampler-has-a-slightly-different-sound-its-too-much-of-a-faff> [Accessed 28 April, 2025]

Future Music. (2015). The Black Dog on building their own MIDI controller and over two decades of making music. *MusicRadar*. Available from: <https://www.musicradar.com/news/tech/the-black-dog-on-building-their-own-midi-controller-and-over-two-decades-of-making-music-629365> [Accessed 28 April, 2025]

Gibson, D. (1993). *Solitudes: Exploring Nature with Music Solitudes*. B000001CZN.

Gillham, A. (2019). Plaid Talk Software, Politics and Where They Find Their Melodies. *Inverted Audio*. Available from: <https://inverted-audio.com/feature/plaid-talk-software-politics-and-where-they-find-their-melodies/> [Accessed 28 April, 2025]

Gold, H. (1967). Vladimir Nabokov, The Art of FictionNo. 40. *The Paris Review*. Available from: <https://www.theparisreview.org/interviews/4310/ the-art-of-fiction-no-40-vladimir-nabokov> [Accessed 28 April, 2025]

Goldmann, S. (2015). *Presets - digital shortcuts to sound*. The Bookworm.

Gonsher, A. (2013). Interview: The Field on Searching for the Perfect Loop. *Red Bull Music Academy*. Available from: <https://daily.redbullmusicacademy.com/2013/10/the-field-interview> [Accessed 28 April, 2025]

Google DeepMind. (2023). Transforming the future of music creation. *Google DeepMind*. Available at: <https://deepmind.google/discover/blog/transforming-the-future-of-music-creation/> [Accessed 28 April, 2025]

Gourley, B. (1999). Anne Dudley and Paul Morley Discuss the Art of Noise. *Chaos Control Digizine*. Available from: <https://chaoscontrol.com/the-art-of-noise/> [Accessed 28 April, 2025]

Graham, D. (2016). Finding the Magic: The Secrets of the Music Producer Daniel Lanois. *The Atlantic*. Available from: <https://www.theatlantic.com/entertainment/archive/2016/05/daniel-lanois-interviewmoogfest-bob-dylan-us-brian-eno/484421/> [Accessed 28 April, 2025]

Graham, P. (2023). How to Do Great Work. *paulgraham.com*. Available from: <https://paulgraham.com/greatwork.html> [Accessed 28 April, 2025]

Graham, P. (2023a). How To Get New Ideas. *paulgraham.com*. Available from: <https://paulgraham.com/getideas.html> [Accessed 28 April, 2025]

Graham, P. (2023b). Superlinear Returns. *paulgraham.com*. Available from: <https://paulgraham.com/superlinear.html> [Accessed 28 April, 2025]

Graham, P. (2004). *Hackers & Painters: Big Ideas from the Computer Age*. Sebastopol, CA, O'Reilly Media.

Graham, P. (2003). Hackers and Painters. *paulgraham.com*. Available from: <https://www.paulgraham.com/hp.html> [Accessed 28 April, 2025]

Greeves, D. (2012). King Creosote & Jon Hopkins: Recording Diamond Mine. *Sound On Sound*. Available from: <https://www.soundonsound.com/people/king-creosote-jon-hopkins-recording-diamond-mine> [Accessed 28 April, 2025]

Hall, J. (2017). In the Field: Klara Lewis Talks Making Music out of Nature. *Red Bull Music Academy*. Available from: <https://www.redbull.com/gb-en/field-recording-klara-lewis-interview> [Accessed 28 April, 2025]

Hall, S.G. (1994). New Age Music: An Analysis of an Ecstasy. *Popular Music and Society*, Vol. 18, No. 2: 23–33. Available from: <doi:10.1080/03007769408591552> [Accessed 28 April, 2025]

Hanging out with Audiophiles podcast. (2025). Episode 138, Cristobal Tapia de Veer. *Hanging Out With Audiophiles*. Available from: <https://podcasts.apple.com/us/podcast/hanging-out-with-audiophiles/id1279148418?i=1000706242735> [Accessed 6 May, 2025]

Hanging out with Audiophiles podcast. (2024). Episode 129, Taylor Deupree & Joseph Branciforte .' *Hanging Out With Audiophiles*. Available from: <https://podcasts.apple.com/us/podcast/howa-ep129-taylor-deupree-joseph-branciforte/id1279148418?i=1000671959632> [Accessed 28 April, 2025]

Hanging out with Audiophiles podcast. (2023). Episode 112, Jon Hopkins. *Hanging Out With Audiophiles*. Available from: <https://podcasts.apple.com/us/podcast/hanging-out-with-audiophiles/id1279148418?i=1000625934786> [Accessed 28 April, 2025]

Hanging out with Audiophiles podcast. (2023a). Episode 116, Clark. *Hanging Out With Audiophiles*. Available from: <https://podcasts.apple.com/us/podcast/hanging-out-with-audiophiles/id1279148418?i=1000638348727> [Accessed 28 April, 2025]

Hanging out with Audiophiles podcast. (2021). Episode 78, Four Tet. *Hanging Out With Audiophiles*. Available from: <https://podcasts.apple.com/ us/podcast/hanging-out-with-audiophiles/id1279148418?i=1000503472234> [Accessed 28 April, 2025]

Harman, C. 1968. How Electronic Music Got That Way. *The Atlantic*. Available from: <https://www.theatlantic.com/magazine/archive/1968/12/how-electronic-music-got-that-way/659405/> [Accessed 28 April, 2025]

Harper, A. 2025. What was Lo-Fi? *Popular Music and Society*. Available from: <doi:10.1080/03007766.2025.2490447> [Accessed 28 April, 2025]

Hartenberger, R, ed. (2016). *The Cambridge Companion to Percussion*. Cambridge, Cambridge University Press.

Headphone Commute. (2024). In the studio with Bogdan Raczynski." *Headphone Commute*. Available from: <https://headphonecommute.com/2024/11/18/in-the-studio-with-bogdan-raczynski/> [Accessed 28 April, 2025]

Headphone Commute. (2023). In the studio with µ-Ziq. *Headphone Commute*. Available from: <https://headphonecommute.com/2023/03/07/in-the-studio-with-µ%E2%80%91ziq/> [Accessed 28 April, 2025]

Headphone Commute. (2023a). In the studio with John Hayes. *Headphone Commute*. Available from: <https://headphonecommute.com/2023/04/26/in-the-studio-with-john-hayes/> [Accessed 28 April, 2025]

Headphone Commute. (2023b). In the studio with Hania Rani. *Headphone Commute*. Available from: <https://headphonecommute.com/2023/09/13/in-the-studio-with-hania-rani/> [Accessed 28 April, 2025]

Headphone Commute. (2023c). In the studio with Album Leaf. *Headphone Commute*. Available from: <https://headphonecommute.com/2023/05/02/in-the-studio-with-album-leaf/> [Accessed 28 April, 2025]

Headphone Commute. (2022). In the Studio with Robot Koch. *Headphone Commute*. Available from: <https://headphonecommute.com/2022/07/06/in-the-studio-with-robot-koch/> [Accessed 28 April, 2025]

Headphone Commute. (2022a). In the Studio with Halina Rice. *Headphone Commute*. Available from: <https://headphonecommute.com/2022/09/28/in-the-studio-with-halina-rice/> [Accessed 27 April 2025]

Headphone Commute. (2021). In the studio with Neil Cowley. *Headphone Commute*. Available from: <https://headphonecommute.com/2021/06/08/in-the-studio-with-neil-cowley/> [Accessed 28 April, 2025]

Headphone Commute. (2021a). In the studio with Biosphere. *Headphone Commute*. Available from: <https://headphonecommute.com/2021/01/25/in-the-studio-with-biosphere/> [Accessed 28 April, 2025]

Headphone Commute. (2019). In the Studio with Jogging House. *Headphone Commute*. Available from: <https://headphonecommute.com/2019/10/03/in-the-studio-with-jogging-house/> [Accessed 28 April, 2025]

Headphone Commute. (2018). In the Studio with Ital Tek. *Headphone Commute*. Available from: <https://headphonecommute.com/2018/09/20/in-the-studio-with-ital-tek/> [Accessed 28 April, 2025]

Headphone Commute. (2016). In the studio with Grischa Lichtenberger. *Headphone Commute*. Available from: <https://headphonecommute.com/2016/01/25/in-the-studio-with-grischa-lichtenberger/> [Accessed 28 April, 2025]

Headphone Commute. (2014). In the studio with Celer. *Headphone Commute*. Available from: <https://headphonecommute.com/2014/10/06/in-the-studio-with-celer/> [Accessed 28 April, 2025]

Headphone Commute. (2014a). In the studio with Richard Devine. *Headphone Commute*. Available from: <https://headphonecommute.com/2014/09/03/in-the-studio-with-richard-devine/> [Accessed 28 April, 2025]

Heater, B. (2013). Switched on Bach: David Cope's computer compositions. *Engadget*. Available from: <https://www.engadget.com/2013-05-28-david-cope.html> [Accessed 28 April, 2025]

Heavyocity Media. (n.d.) Interview: Cliff Martinez - Music Composer. *Heavyocity*. Available from: <https://heavyocity.com/artist/cliff-martinez/> [Accessed 28 April, 2025]

Hebdige, Dick. (1987). *Cut 'N' Mix: Culture, Identity and Caribbean Music*. London, Routledge.

Herlihy, G. (2021). Four Tet/Kieran Hebden. *Sound On Sound*. Available from: <https://www.soundonsound.com/people/four-tet-kieran-hebden?amp> [Accessed 28 April, 2025]

Herndon, H. (2021). Holly Plus. Available from: <https://holly.mirror.xyz/54ds2IiOnvthjGFkokFCoaI4EabytH9xjAYy1irHy94> [Accessed 28 April, 2025]

Hertz, T.J. (2020). Producers. Available from: <https://x.com/keinobjekt/status/1216390516054601733?%20lang=en> [Accessed 28 April, 2025]

Hiatt, B. (2023). Ready to Sing Elvis Karaoke…as Elvis? The Weird Rise of AI Music. Rolling Stone. Available from: <https://www.rollingstone.com/music/music-features/ai-music-drake-weeknd-ghostwriter977-beatles-elvis-voice-1234770094/> [Accessed 28 April, 2025]

Hillier, P. (1997). *Arvo Pärt (Oxford Studies of Composers)*. Oxford, Oxford University Press.

Hochberg, W. (2014). When Robots Write Songs. *The Atlantic*. Available from: <https://www.theatlantic.com/entertainment/archive/2014/08/computers-that-compose/374916/> [Accessed 28 April, 2025]

Hogwood, B. (2024). Interview: Loscil – Granular Processes, Emotive Results. *Arcana.Fm.* Available from: <https://arcana.fm/2016/12/18/loscil/> [Accessed 28 April, 2025]

Hood, M. (1959). The Challenge of 'Bi-Musicality'. *Ethnomusicology*, Vol. 4, No. 2: 55–59.

Hopkins, J. (2021). I am Jon Hopkins - Ask Me Anything! *Reddit.* Available from: <https://www.reddit.com/r/electronicmusic/comments/mw6yp5/i_am_jon_hopkins_ask_me_anything/> [Accessed 28 April, 2025]

Housego, N. (2020). Interview: KMRU. *Mezzanine Mastering.* Available from: <https://mezzaninemastering.com/interview-kmru> [Accessed 28 April, 2025]

Howells, T. (2016). Biosphere Comes in from the Cold on New Album Departed Glories. *Fact Magazine.* Available from: <https://www.factmag.com/2016/09/27/biosphere-interview-departed-glories/> [Accessed 28 April, 2025]

Inglis, S. (2024). Dune: Hans Zimmer & Friends. *Sound On Sound.* Available from: <https://www.soundonsound.com/techniques/dune-hans-zimmer-friends> [Accessed 28 April, 2025]

Inglis, S. (2001). Guy Sigsworth: Music After MIDI. *Sound On Sound.* Available from: <https://www.soundonsound.com/people/guy-sigsworth-music-after-midi> [Accessed 28 April, 2025]

Inner Islands. (n.d.) Interview with Dylan Henner. *Inner Islands.* Available from: <https://inner-islands.tumblr.com/post/614392210293702656/%20interview-with-dylan-henner-1-what-are-some/amp> [Accessed 28 April, 2025]

Jackson, M. (1981). Billie Jean (1981 Home Demo). *YouTube.* Available from: <https://www.youtube.com/watch?v=2Hp3P4XOT_U> [Accessed 28 April, 2025]

Jenkins, D. (2020). A Rare Audience with Dillinja. *UKF.* Available from: <https://ukf.com/read/a-rare-audience-with-dillinja/> [Accessed 28 April, 2025]

Joyce, J. (2024). Download the Live Set of Polygonia's New Track 'Neon Oracle' . *Ableton.* Available from: <https://www.ableton.com/en/blog/download-the-live-set-of-polygonias-new-track-neon-oracle/> [Accessed 28 April, 2025]

Kamer, F. (2022). How Fred Again. Turns Digital Bricolage into Dance-Floor Weepers. *The New York Times.* Available from: <https://www.nytimes.com/2022/10/26/arts/music/fred-again-actual-life.html> [Accessed 28 April, 2025]

Keil, Charles. (1987). Participatory Discrepancies and the Power of Music. *Cultural Anthropology*, Vol. 2, No. 3: 275–283.

Keller, D. (2007). Bruce Springsteen's 'Nebraska' - A PortaStudio, two SM57's, and Inspiration. *Tascam.* Available from: <https://tascam.jp/int/support/news/481> [Accessed 28 April, 2025]

Kelly, K. (2023). *Excellent Advice for Living.* New York, Viking.

Kelly, K. (1995). Gossip Is Philosophy. *Wired.* Available from: <https://www.wired.com/1995/05/eno-2/> [Accessed 28 April, 2025]

Kelly, K. and S.Johnson. (2018). Kevin Kelly and Steven Johnson on Where Ideas Come From. *Wired.* Available from: <https://www.wired.com/2010/09/mf-kellyjohnson/> [Accessed 28 April, 2025]

Klinkenborg, V. (2012). *Several Short Sentences About Writing.* New York, Vintage.

Kocienda, K. (2018). *Creative Selection: Inside Apple's Design Process During the Golden Age of Steve Jobs.* New York, St. Martin's.

Koren, L. (2023). *On Creating Things Aesthetic.* Point Reyes, CA, Imperfect Publishing.

Koren, L. (2008). *Wabi-Sabi for Artists, Designers, Poets & Philosophers.* Point Reyes, CA, Imperfect Publishing.

König, B.R. (2019). Floating Points: Most of the time, I'm on Discogs. *Groove.* Available from: <https://www.groove.de/2019/10/18/floating-points-most-of-the-time-im-on-discogs/> [Accessed 28 April, 2025]

Kraker, L. (2018). Ableton Master G Jones Breaks down His Creative Approach on the Ineffable Truth EDM. Available from: <https://edm.com/interviews/g-jones-interview> [Accessed 28 April, 2025]

Krasowski, W. (2008). Infinitive Thing: Interview with Brian McBride of Stars of the Lid. *Infinitive Thing.* Available from: <https://infinitivething.blogspot.com/2008/07/interview-with-brian-mcbride-from-stars.html> [Accessed 28 April, 2025]

Kraus, R. (1979). Grids. October, Vol. 9 (Summer): 50–64. *Sound On Sound Electronic Music podcast.* (2024). *Krust—Drum & Bass Pioneer.* Available from: <https://podcasts.apple.com/us/podcast/electronic-music/id1512947854?i=1000675183906> [Accessed 28 April, 2025]

Kubik, Gerhard. (1962). The Phenomenon of Inherent Rhythms in East and Central African Instrumental Music. *African Music*, Vol. 3, No. 1: 33–42.

Kuprenas, J. and M. Frederick. (2018). *101 Things I Learned in Engineering School.* New York, Crown.

Lakoff, G. and M. Johnson. (2008). *Metaphors We Live By.* Chicago, University of Chicago Press.

Lanier, J. (2023). What My Musical Instruments Have Taught Me. *The New Yorker.* Available from: <https://www.newyorker.com/culture/the-weekend-essay/what-my-musical-instruments-have-taught-me> [Accessed 28 April, 2025]

Lanza, J. (2004). *Elevator Music: A Surreal History of Muzak, Easy-Listening, and Other Moodsong.* Ann Arbor, MI, University of Michigan Press.

Lateanonymous474. (2025). Maybe that's just me, but the samples sound pretty basic. *Output Co-Producer|Quick Walkthrough.* Available from <https://www.youtube.com/watch?v=NSfearFMLQY> [Accessed 28 April, 2025]

Lee, S. (2023). Shima: Full Automony Over Everything. *Ableton.* Available from: <https://www.ableton.com/zh-cn/blog/shima-full-autonomy-over-everything/> [Accessed 28 April, 2025]

Lichterwaldt, D. (2024). Interview with Lucy Railton. *Les Nouveaux Riches.* Available from: <https://www.les-nouveaux-riches.com/interview-with-lucy-railton/> [Accessed 28 April, 2025]

Lidström, F. (2024). Synplant User Guide. *Sonic Charge.* Available from: <https://soniccharge.com/public/Synplant%20User%20Guide.pdf?dl=20240830T095912&inline> [Accessed 28 April, 2025]

Liu, L. et al. (2021). Understanding the onset of hot streaks across artistic, cultural, and scientific careers. *Nature Communications.* Available from: <https://www.nature.com/articles/ s41467–41021–25477–25478> [Accessed 28 April, 2025]

Live Producers Online. (2023). Interview with Frequent - Ableton Music Producer Podcast: Episode 150. Available from: <https://www.youtube.com/watcv=KSHpqZbtWfA> [Accessed 28 April, 2025]

Lockie, A. (2021). Finding The 'And' of a Musical World: An Interview with Rachika Nayar. *Slug Mag.* Available from: <https://www.slugmag.com/music/interviews/music-interviews/an-interview-with-rachika-nayar/> [Accessed 28 April, 2025]

Loughridge, D. (2023). *Sounding human: music and machines.* 1740/2020. Chicago, University of Chicago Press.

Lowe, Z. (2022). Fred Again.: 'Actual Life 3', Boiler Room, and Creative Process. *Apple Music.* Available from: <https://www.youtube.com/watch?v=zeD0g5xXo7E> [Accessed 28 April, 2025]

Luhmann, N. (1981). Communicating with Slip Boxes. Available from: <https://luhmann.surge.sh/communicating-with-slip-boxes> [Accessed 28 April, 2025]

Lynch, D. (2007). *Catching the Big Fish.* New York, TarcherPerigee.

Lysaker, J.T. (2018). *Brian Eno's Ambient 1: Music for Airports (The Oxford Keynotes Series).* Oxford, Oxford University Press.

Mackintosh, H. (2022). Proc Fiskal. *MusicRadar.* Available from: <https://www.musicradar.com/news/proc-fiskal> [Accessed 28 April, 2025].

Mackintosh, H. (2021). Jon Hopkins: 'I Don't Really Have Any Modular Synths, Drum Machines or Any of That Stuff - It's All Really Ableton and Source Audio from Random Places .' *MusicRadar.* Available from: <https://www.musicradar.com/news/jon-hopkins-modular-synths-drum-machines-ableton> [Accessed 28 April, 2025]

Mandelbrot, Benoit B. (1983). *The Fractal Geometry of Nature.* San Francisco, CA, W.H. Freeman and Company.

Marchese, D. (2022). Brian Eno Reveals the Hidden Purpose of All Art. *The New York Times.* Available from: <https://www.nytimes.com/interactive/2022/11/14/magazine/brian-eno-interview.html0> [Accessed 28 April, 2025]

Marlow, O. (2013). Sonic Router 030: EPROM's Fundamental Particles. *The Quietus*. Available from: <https://thequietus.com/interviews/sonic-router-030-eprom-interview-halflife/> [Accessed 28 April, 2025]

Maurer, J.A. (1999). A Brief History of Algorithmic Composition. Available from: <https://ccrma.stanford.edu/~blackrse/algorithm.html> [Accessed 28 April, 2025]

McClean, C. (2023). An Interview with Perila. *Subbacultcha*. Available from: <https://subbacultcha.nl/2023/05/05/an-interview-with-perila/> [Accessed 28 April, 2025]

McEvoy, C. (2024). I'm Not Excited about AI Making Stuff That a Human Could Do . *MusicRadar*. Available from:v<https://www.musicradar.com/news/imogen-heap-interview> [Accessed 28 April, 2025]

McEvoy, C. (2022). Ghost in the Drum Machine: How Creative AI Is Kicking off a Paradigm Shift in Music. *MusicTech*. Available from: <https://musictech.com/features/opinion-analysis/ai-artist-developers-isongwriting-production/> [Accessed 28 April, 2025]

McKinney, M. (2023). An Interview with Lee Gamble. *Passion of the Weiss*. Available from: <https://www.passionweiss.com/2023/10/18/lee-gamble-interview-models/> [Accessed 28 April, 2025]

McNeil, M. (2022). Ropes, Fields and Waves: Celia Hollander Sculpts Time with Sound. *Ableton*. Available from: <https://www.ableton.com/en/blog/ropes-fields-and-waves-celia-hollander-sculptstime-with-sound> [Accessed 28 April, 2025].

McPhee, J. (2022). Tabula Rasa: Volume Three. *The New Yorker*. Available from: <https://www.newyorker.com/magazine/2022/02/07/tabula-rasa-volume-three> [Accessed 28 April, 2025]

McTernan, C. (2023). The Art of Production: Carmen Villain. *Resident Advisor*. Available from: <https://ra.co/features/4155> [Accessed 28 April, 2025]

Miller, C. (2014). Little White Earbuds Interviews Huerco S. *Little White Earbuds*. Available from: <http://www.littlewhiteearbuds.com/feature/little-white-earbuds-interviews-huerco-s/> [Accessed 28 April, 2025]

Moller, M.Y. (2021). Colleen - See Silly Shot's Synth Sounds. *Martin Yam Moller*. Available from: <https://martinyammoller.com/9oddquestionsformusicgearjunkies/colleen-see-silly-shots-synth-sounds/> [Accessed 28 April, 2025]

Moller, M.Y. (2020). Anna-Karin Berglund - Deeply Diving Droner. *Martin Yam Moller*. Available from: <https://martinyammoller.com/9oddquestionsformusicgearjunkies/anna-karin-berglund-deeply-diving-droner/> [Accessed 28 April, 2025]

Monroe, J. (2023). Grimes Unveils Software to Mimic Her Voice, Offering 50–50 Royalties for Commercial Use. *Pitchfork*. Available from: <https://pitchfork.com/news/grimes-unveils-software-to-mimic-her-voice-and-announces-2-new-songs/> [Accessed 28 April, 2025]

Monroe, J. (2022). Nala Sinephro Is Subverting Jazz Tradition One Immaculate Note at a Time. *Pitchfork*. Available from: <https://pitchfork.com/features/rising/nala-sinephro-interview/> [Accessed 28 April, 2025]

Moorefield, V. (2010). *The Producer as Composer: Shaping the Sounds of Popular Music*. Cambridge, MIT Press.

Moreno, V. (2024). Autechre: Immersive Listening, Synesthesia, and the Manchester Scene. *Metal Magazine*. Available from: <https://metalmagazine.eu/en/post/autechre> [Accessed 28 April, 2025]

Mori, M. (2012). The Uncanny Valley: The Original Essay by Masahiro Mori." *IEEE Spectrum*. Available from: <https://spectrum.ieee.org/the-uncanny-valley> [Accessed 28 April, 2025]

Moss, R. (2019). Interview: Proc Fiscal. *The Art of Grime*. Available from: <https://theartofgrime.com/2019/09/12/interview-proc-fiskal/> [Accessed 28 April, 2025]

mrandish. (2024). Udio – AI Music Generator. *Hacker News*. Available from: <https://news.ycombinator.com/item?id=40557981> [Accessed 28 April, 2025]

Muggs, J. (2016). Autechre: elseq et al. *Resident Advisor*. Available from: <https://ra.co/features/2756> [Accessed 28 April, 2025]

Mullen, M. (2024). Kiasmos on the 'Perfect Mono Synth: 'It Doesn't Need Anything on the Chain and It Still Sounds Great, Which You Can't Say about Most Synths . *MusicRadar*. Available from: <https://www.musicradar.com/news/kiasmos-interview> [Accessed 28 April, 2025]

Mullen. M. (2023). James Holden. *Music Radar*. Available from: <https://www.musicradar.com/news/james-holden-i-got-really-bored-of-making-music-but-the-moment-i-discovered-modular-and-live-and-maxmsp-it-all-opened-up-again> [Accessed 28 April, 2025]

Mullen, M. (2023a). Nathan Micay: 'Less Is Not More. If I Can Make Ten Different Weird Sounds Work Together, That's Awesome . *MusicRadar*. Available from: <https://www.musicradar.com/news/nathan-micay-interview> [Accessed 28 April, 2025]

Mullen, M. (2023b). DJ Shadow on Why He'll Never Go Back to the MPC. *MusicRadar*. Available from: <https://www.musicradar.com/news/dj-shadow-on-why-hell-never-go-back-to-the-mpc-i-got-the-newest-mpc-and-i-lasted-about-a-day-and-a-half-when-you-know-what-the-possibilities-are-now-and-you-try-to-go-back-in-time-it-doesnt-work> [Accessed 28 April, 2025]

Mullen, M. (2023c). Produce Like... Model Man: 'Ableton Was Just the Thing I Needed to Connect the Dots between Classical Piano and Production.' *MusicRadar*. Available from: <https://www.musicradar.com/news/produce-like-model-man> [Accessed 28 April, 2025]

Mullen, M. (2023d). Lucinda Chua: In My Production, I'm Just Trying to Carve out a Space Where I Feel like I Belong. *MusicRadar*. Available from: <https://www.musicradar.com/news/lucinda-chua-interview> [Accessed 28 April, 2025]

Mullen, M. (2023e). How Patten Used Text-to-Audio AI to Make an Entire Album: 'We're at the Precipice of a Fundamental Shift in How We Think about Making Music.' *MusicRadar*. Available from: <https://www.musicradar.com/news/patten-interview> [Accessed 28 April, 2025]

Mullen, M. (2023f). 'Vocal Deepfakes Are Here to Stay': AI Voice Cloning Is about to Change Pop Music Forever. Here's Why. *MusicRadar*. Available from: <https://www.musicradar.com/news/ai-voice-cloning> [Accessed 28 April, 2025]

Mullen, M. (2022). Loraine James on How Slate + Ash Plugins and the Hologram Electronics Microcosm Inspired the Sound of Whatever the Weather. *MusicRadar*. Available from: <https://www.musicradar.com/news/loraine-james-whatever-the-weather-interview> [Accessed 28 April, 2025]

Mullen, M. (2022a). Nils Fram's analogue spirituality. *MusicRadar*. Available from: <https://www.musicradar.com/news/nils-frahm-analogue-spirituality> [Accessed 28 April, 2025]

Mullen, M. (2022b). Tom VR: 'I've Listed the Korg ARP Odyssey on Gumtree Three Times, but I Always Come Back to It.' *MusicRadar*. Available from: <https://www.musicradar.com/news/tom-vr-interview> [Accessed 28 April, 2025]

Mullen, M. (2021). Nathan Fake: 'Drums Are Usually My Starting Point. Despite My Music Being Quite Melodic, I've Always Been Obsessed with Electronic Drums and Abstract Percussion.' *MusicRadar*. Available from: <https://www.musicradar.com/news/nathan-fake-interview> [Accessed 28 April, 2025]

Murray, R. (2018). Everything is Going to Be Alright: Yamaneko Interviewed. *Clash Music*. Available from: <https://www.clashmusic.com/features/everything-is-going-to-be-alright-yamaneko-interviewed/> [Accessed 28 April, 2025]

MusicRadar Tech. (2025). Djrum in the Studio: Making Frekm Pt.2. *MusicRadar Tech*. Available from: <https://www.youtube.com/watch?v=zyMDZZ8-qKg> [Accessed 28 April, 2025]

MusicRadar. (2022). Kelly Lee Owens: 'Analogue Keeps Things Interesting. It Rebels against Stability.' *MusicRadar*. Available from: <https://www.musicradar.com/news/kelly-lee-owens-interview> [Accessed 28 April, 2025]

MusicRadar Tech. (2019). DLR Breaks down the Creation of Trick – The Track. *MusicRadar Tech*. Available from: <https://www.youtube.com/watch?v=2VYqaYi0nhY> [Accessed 28 April, 2025]

My Forever Studio podcast. (2023). Amon Tobin, Episode 53. Available from: <https://podcasts.apple.com/us/podcast/ep-53-amon-tobins-floating-chamber-of-synthesis/id1494851400?i=1000607850825> [Accessed 28 April, 2025]

Nagle, P. (2015). Xfer Records Serum. *Sound On Sound*. Available from: <https://www.soundonsound.com/reviews/xfer-records-serum> [Accessed 28 April, 2025]

Nair, V. (2013). An Interview with Nils Frahm. *Designing Sound*. Available from: <https://designingsound.org/2013/03/27/an-interview-with-nils-frahm/> [Accessed 28 April, 2025]

Napolitano, P. (2010). The Silences of Biosphere. Neo Baroque and Aware Intimacy. *Digicult*. Available from: <https://digicult.it/digimag/issue-056/the-silences-of-biosphere-neo-baroque-and-aware-intimacy/> [Accessed 28 April, 2025]

Nelson, H. (2014). 'I'm So Minimal, I'm Not Even Minimalist!': Harold Budd Interviewed. *The Quietus*. Available from: <https://thequietus.com/interviews/harold-budd-interview/> [Accessed 28 April, 2025]

New Bliss. (2019). Royal Library | Rain and Thunderstorm Sounds on Study Ambience with Crackling Fireplace. *New Bliss*. Available from: <https://www.youtube.com/watch?v=CHFif_y2TyM> [Accessed 28 April, 2025]

Newport, C. (2023). What Kind of Mind Does ChatGPT Have? *The New Yorker*. Available from: <https://www.newyorker.com/science/annals-of-artificial-> [Accessed 28 April, 2025]

Nicholas, G. (2014). Q&A: Clark Interview (UK, WARP) With George Nicholas. *Live School*. Available from: <https://blog.liveschool.net/clark-interview-warp-uk-interview-george-seekae/> [Accessed 28 April, 2025]

Noah, P. (2024). Modulation Revelation: The Future of Sound Design? *Attack Magazine*. Available from: <https://www.attackmagazine.com/features/long-read/modulation-revelation-the-future-of-sound-design/> [Accessed 28 April, 2025]

Noë, A. (2015). What Art Unveils. *The New York Times*. Available from: <https://deref-gmx.com/mail/client/AMD8YAnAZVk/dereferrer/?redirectUrl=https%3A%2F%2Farchive.nytimes.com%2Fopinionator.blogs.nytimes.com%2F2015%2F10%2F05%2Fwhat-art-unveils%2F%3Fsmid%3Dtw-share%26referer%3Dhttp%3A%2F%2Ft.co%2FUo5hzC0Mls" \t "_blank"https://archive.nytimes.com/opinionator.blogs.nytimes.com/2015/10/05/what-art-unveils/?smid=tw-share&referer=http://t.co/Uo5hzC0Mls> [Accessed: 14 July 2025]

Nosnitsky, A. (2018). Aphex Twin's Mask Collapses. *Crack Magazine*. Available from: <https://crackmagazine.net/article/long-reads/aphex-twins-mask-collapses/> [Accessed 28 April, 2025]

Obenour, D.C. (2023). Colleen's Cécile Schott. *Off Shelf*. Available from: <https://offshelf.net/2023/09/12/colleens-cecile-schott/> [Accessed 28 April, 2025]

Odell, J. (2019). *How To Do Nothing: Resisting the Attention Economy*. New York, Melville House.

Oliveros, P. (2005). *Deep Listening: A Composer's Sound Practice*. Bloomington, Indiana, iUniverse.

OpenAI. (2020.). *Jukebox*. Available from: <https://openai.com/index/jukebox/> [Accessed 28 April, 2025]

Oram, D. (2016). *Daphne Oram: An Individual Note of Music, Sound and Electronics*. London, Anomie Academic.

Output. (2025). The Fastest Way to Find Samples That Fit. *Output*. Available from: <https://output.com/ products/co-producer> [Accessed 28 April, 2025]

Output. (2024). Describe a vibe to generate custom sample packs. *Output*. Available from: <https://output.com/products/pack-generator> [Accessed 28 April, 2025]

Pareles, J. (2022). Autechre Worked in Isolation for Decades. Now It's Unintentionally Timely. *The New York Times*. Available from: <https://www.nytimes.com/2020/10/13/arts/music/autechre-signinterview.html> [Accessed 28 April, 2025]

Pareles, J. (2013). Brothers Who Make Electronica by Hand. *The New York Times*. Available from: <https://www.nytimes.com/2013/06/12/arts/music/tomorrows-harvest-by-boards-of-canada.Html> [Accessed 28 April, 2025]

Passard, A. (2016). The Gesture – Chef Alain Passard. *The Hungry Philosopher*. Available at: <https://hungryphil.com/2016/09/09/the-gesture-alain-passard/> [Accessed: 28 April, 2025].

Patel, D. (2025). *The Scaling Era: An Oral History of AI, 2019–2025*. San Francisco, CA, Stripe Press.

Patel, N. (2025). Splice CEO Kakul Srivastava on why push-button AI is 'insulting' to musicians. *Decoder* podcast. Available from: <https://podcasts.apple.com/us/podcast/decoder-with-nilay-patel/id1011668648?i=1000700572405> [Accessed 28 April, 2025]

PaulXStretch. (n.d.). Extreme Time-Stretching. *Sonosaurus*. Available from: <https://sonosaurus.com/paulxstretch/> [Accessed 28 April, 2025]

Payne, O. (2023). 'Ambient Music Is about Drawing out the Details in the Small Sounds': Producer Cephas Azariah on His Essential Reverb Plugins. *MusicTech*. Available from: <https://musictech.com/features/interviews/cephas-azariah-if-i-can-be-honest-anjunadeep-reflections/> [Accessed 28 April, 2025]

Pemberton, N.T. (2019). Electronic Producer Actress Wants to Know How an AI Painted a Face. *Interview Magazine*. Available from: <https://www.interviewmagazine.com/music/electronic-producer-actress-wants-to-know-how-an-artificial-intelligence-painted-a-face> [Accessed 28 April, 2025]

Pensado, D. (2021). Interview with Dave Pensado. *Gearspace*. Available from: <https://gearspace.com/board/interviews/1364950-interview-dave-pensado.html> [Accessed 28 April, 2025]

Petrusich, A. (2023). The Emotionally Haunted Electronic Music of Oneohtrix Point Never. *The New Yorker*. Available from: <https://www.newyorker.com/magazine/2023/10/02/the-emotionally-haunted-electronic-music-of-oneohtrix-point-never> [Accessed 28 April, 2025]

Pirsig, R. (1974). *Zen and the Art of Motorcycle Maintenance*. New York, Bantam Books.

Pitchfork. (2007). The Field. *Pitchfork*. Available from: <https://pitchfork.com/features/interview/6587-the-field/> [Accessed 28 April, 2025]

Polanyi, M. (2009). *The Tacit Dimension*. Chicago, University of Chicago Press.

Prendergast, M. (1986). The Sound-Painted World of Harold Budd. *Sound On Sound*. Available from: <https://www.muzines.co.uk/articles/the-sound-painted-world-of-harold-budd/1448> [Accessed 28 April, 2025]

Prestigious_Gold_742. (2023). How to find your sound? *Reddit*. Available from: <www.reddit.com/r/ableton/comments/138au1b/how_to_find_your_sound/> [Accessed 28 April, 2025]

prestonloyola. (2008). DnB 1996 (LolaDaMusica) Part2: Photek. *YouTube*. Available from: <https://www.youtube.com/watch?v=TyKuzY52hWU>.

Quiet Details. (2024). R Beny Q&A Part Two and Last Day of Half-Price. *Quiet Details*. Available from: <https://quietdetails.substack.com/p/r-beny-q-and-a-part-two-and-last> [Accessed 28 April, 2025]

Ravens, C. (2023). Patten Taps into Text to Audio AI's Musical Potential. *Bandcamp Daily*. Available from: <https://daily.bandcamp.com/features/patten-mirage-fm-interview> [Accessed 28 April, 2025]

Ravens, C. (2020). Gábor Lázár. *Native Instruments*. Available from: <https://blog.native-instruments.com/sketches-gabor-lazar/> [Accessed 28 April, 2025]

Reich, S. (2022). *Conversations*. New York, Hanover Square Press.

Remer, T.G., ed. (1965) *Serendipity and the Three Princes*. Normal, University of Oklahoma Press.

RA Exchange podcast. (2024). Lee Gamble, Episode 729. *RA Exchange*. Available from: <https://podcasts.apple.com/us/podcast/ex-729-lee-gamble/id395896252?i=1000668471662> [Accessed 28 April, 2025]

RA Exchange podcast. (2023). Moritz von Oswald, Episode 690. *RA Exchange*. Available from: <https://podcasts.apple.com/us/podcast/ex-690-moritz-von-oswald/id395896252?i=1000637802009> [Accessed 28 April, 2025]

RA Exchange podcast. (2022). Clark, Episode 823. *RA Exchange*. Available from: <https://ra.co/podcast/823> [Accessed 28 April, 2025]

Reuter, Anders. (2022). Who Let the DAWs Out? The Digital in a New Generation of the Digital Audio Workstation. *Popular Music and Society*, Vol. 45, No. 2: 113–128. Available from: <doi:10.1080/03007766.2021.1972701> [Accessed 28 April, 2025]

Reynaldo, S. (2025). What Is Carrier All About? *Substack*. Available from: <https://firstfloor.substack.com/p/what-is-carrier-all-about?utm_source=substack&utm_campaign=post_embed&utm_medium=email> [Accessed 28 April, 2025]

Reynolds, S. (2011). *Retromania: Pop Culture's Addiction to its Own Past*. New York, Farrar, Straus and Giroux.

Ridley, M. (2021). *How Innovation Works*. New York, Harper.

Roads, C. (2004). *Microsound*. Cambridge, MIT Press.

Roberts, L. (2022). In Conversation: Koreless. *Clash*. Available from: <https://www.clashmusic.com/features/in-conversation-koreless/> [Accessed 28 April, 2025]

Robina, W.T. (2021). *Technical manifesto for the deviant sound engineer*. Concord, MA, aPriori press.

Roche, S. (2025.) Output's new Co-Producer plugin uses AI to curate the perfect samples for your project – all from within your DAW. *MusicTech*. Available from: <HYPERLINK "https://musictech.com/news/gear/output-co-producer/" \t "_blank"https://musictech.com/news/gear/output-co-producer/> [Accessed 10 July, 2025]

Rolling Stone. (2024). Soul of the Machine. *Soundcloud*. Available from: <https://soundcloud.com/rs-539916550/soul-of-the-machine> [Accessed 28 April, 2025]

Roose, K. (2022). An A.I.-Generated Picture Won an Art Prize. Artists Aren't Happy. *The New York Times*. Available from: <https://www.nytimes.com/2022/09/02/technology/ai-artificial-intelligence-artists.html> [Accessed 28 April, 2025]

Rose, B. (2023). Christina Vantzou's Worlds of Sound. *Foxy Digitalis*. Available from: <https://foxydigitalis.zone/2023/01/18/christina-vantzous-worlds-of-sound/> [Accessed 28 April, 2025]

Ross, G. (2021). How Fred Again. Hit the Big Time by Turning His Life into a Symphony. *Mixmag*. Available from: <https://mixmag.net/feature/fred-again-cover-feature-interview> [Accessed 28 April, 2025]

Rothman, J. (2025). Why Even Try If You Have A.I.? *The New Yorker*. Available from: <https://www.newyorker.com/culture/open-questions/why-even-try-if-you-have-ai> [Accessed 28 April, 2025]

Rubik, E. (2020). *Cubed: The Puzzle of Us All*. New York, Flatiron Books.

Rutherford-Johnson, T. (2013). The Influence Engine: Steve Reich and Pop Music. *New Music USA*. Available from: <https://newmusicusa.org/nmbx/the-influence-engine-steve-reich-and-pop-music/> [Accessed 28 April, 2025]

Sannicandro, J. (2019). Sound Propositions 016: Felicia Atkinson. *A Closer Listen*. Available from: <https://acloserlisten.com/2019/07/05/sound-propositions-016-felicia-atkinson/> [Accessed 28 April, 2025]

Sannicandro, J. (2017). Sound Propositions 011: Taylor Deupree. *Sound Propositions*. Available from: <https://soundpropositions.com/2017/08/18/sound-propositions-011-taylor-deupree/> [Accessed 28 April, 2025]

Scarth, G. (2013). James Holden. *Attack Magazine*. Available from: <https://www.attackmagazine.com/features/interview/james-holden/> [Accessed 28 April, 2025]

Scarth, G. and R. Linn. (2013). Roger Linn On Swing, Groove & The Magic of the MPC's Timing. *Attack Magazine*. Available from: <https://www.attackmagazine.com/features/interview/roger-linn-swing-groove-magic-mpc-timing/> [Accessed 28 April, 2025]

Schafer, R.M. (1977). *The Soundscape: Our Sonic Environment and the Tuning of the World*. Vermont, Destiny Books.

Schmidt, T. (2008). Moritz von Oswald. *Red Bull Music Academy*. Available at: <https://www.redbullmusicacademy.com/lectures/moritz-von-oswald-early-morning-freestyles> [Accessed 28 April, 2025]

Schütze, P. (1995). Strategies for Making Sense. *More Dark Than Shark*. Available at: <https://www.moredarkthanshark.org/eno_int_wire-sep95.html> [Accessed 28 April, 2025]

Seeger, C. (1976). The Musicological Juncture. *Ethnomusicology* 21, no. 2: 179–188. Available at: <doi:10.2307/850942> [Accessed 28 April, 2025]

Shenton A., ed. (2012). *The Cambridge Companion to Arvo Pärt*. Cambridge, Cambridge University Press.

Sherburne, P. (2014). Strange Visitor: A Conversation with Aphex Twin. *Pitchfork*. Available at: <https://pitchfork.com/features/cover-story/9506-strange-visitor-a-conversation-with-aphex-twin/?page=2> [Accessed 28 April, 2025]

Sherrill, M. (2025). New World Symphonies: Listening for the Future of Music. *Harper's Magazine*. Available at: <https://harpers.org/archive/2025/03/new-world-symphonies-matthew-sherrill-future-of-music/> [Accessed 28 April, 2025]

Sigurdarson, B. (2018). Bjarki - Waveshaper TV Ep.2. *Waveshaper Media*. Available at: <https://www.youtube.com/watch?v=QYRxid1iss0> [Accessed 28 April, 2025]

Shopmaker, Z. (2022). Blazing Unexplored Terrain: The Ambient Innovations of Huerco S. *Resident Advisor*. Available at: <https://ra.co/features/3972> [Accessed 28 April, 2025]

Silver, N. (2024). *On The Edge: The Art of Risking Everything*. New York, Penguin Press.

Simpson, D. (2018). Jean-Michel Jarre: How We Made Oxygène. *The Guardian*. Available at: <https://www.theguardian.com/music/2018/oct/16/jean-michel-jarre-michel-.granger-oxygene> [Accessed 28 April, 2025]

Sjöberg, F. (2017). *The Art of Flight*. New York, Penguin Books.

Smith, M. (2020). The Art of Production: Rose. *Resident Advisor*. Available at: <https://ra.co/features/3619> [Accessed 28 April, 2025]

Snape, J. and G. Born. (2022). Max, music software and the mutual mediation of aesthetics and digital technologies. In: Born, G., ed. *Music and Digital Media: A Planetary Anthropology*. London: UCL Press, pp. 220–266.

Sonosaurus. (n.d.). *PaulXStretch*. Available from: <https://sonosaurus.com/paulxstretch/ > [Accessed 28 April, 2025]

SOS Recording & Mixing podcast. (2025). Using Tube and Tape Plug-Ins. Available from: <https://podcasts.apple.com/us/podcast/recording-mixing/id1512947858?i=1000699008088> [Accessed 28 April, 2025]

SOS Electronic Music podcast. (2024). Krust - Drum & Bass Pioneer. Available from: <https://podcasts.apple.com/us/podcast/electronic-music/id1512947854?i=1000675183906> [Accessed 28 April, 2025]

SOS Electronic Music podcast. (2023). Graham Massey - 808 State. Available from: <https://podcasts.apple.com/us/podcast/electronic-music/id1512947854?i=1000601069785> [Accessed 28 April, 2025]

Spice, A. (2015). The Dangers of Digital: Brian Eno on Technology and Modern Music. *The Vinyl Factory*. Available from: <https://thevinylfactory.com/features/the-dangers-of-digital-brian-eno-on-technology-and-modern-music/> [Accessed 28 April, 2025]

Spitfire Audio. (2018). The Music of Chris Clark. *Spitfire Audio*. Available from: <https://www.youtube.com/watch?v=vSCIUTE6Vs4> [Accessed 28 April, 2025]

stability.ai. (2024). *Introducing Stable Audio 2.0*. Available from: <https://stability-ai.squarespace.com/news/stable-audio-2-0> [Accessed 28 April, 2025]

Stebbings, H. (2025). 20VC: The Future of Foundation Models. Available from: <https://thetwentyminutevc.libsyn.com/20vc-the-future-of-foundation-models-the-future-of-ai-consumer-apps-and-why-openai-did-a-disservice-to-them-the-future-of-music-spotify-vs-youtube-spotify-vs-tiktok-what-happens-with-mikey-shulman-suno> [Accessed 28 April, 2025]

Steinberg. (n.d.). *VST Classics Vol. 1*. Available from: <https://o.steinberg.net/en/support/unsupported_software/vst_classics_vol_1_2.html> [Accessed 28 April, 2025]

Stevens, W. (2011). *The Collected Poems of Wallace Stevens*. New York, Vintage International.

Stimming, M. (2016). Stimming Exclusive Interview: 'At a Certain Point, the Music is the Boss.' *Pointblank Music School*. Available from: <https://plus.pointblankmusicschool.com/stimming-at-a-certain-point-the-music-is-the-boss/> [Accessed 28 April, 2025]

Stout, J. (2018). Interview with Robert Dudzic. *Designing Sound*. Available from: <https://designingsound.org/2018/08/09/interview-with-robert-dudzic/> [Accessed 28 April, 2025]

Strange, A. (1983). *Electronic Music: systems, techniques, and controls*. Dubuque, Iowa, Wm. C. Brown Company Publishers.

Sudnow, D. (2001). *Ways of the Hand: a rewritten account*. Cambridge, MA, MIT Press.

Taleb, N. (2012). *Antifragile: Things That Gain from Disorder*. New York, Random House.

Taleb, N. (2012a). *Understanding is a Poor Substitute for Convexity (Antifragility)*. Available from: <https://fooledbyrandomness.com/ConvexityScience.pdf> [Accessed 28 April, 2025]

Taleb, N. (2010). *The Bed of Procrustes: Philosophical and Practical Aphorisms*. New York, Random House.

Tape Notes podcast. (2024). Episode 144, Bonobo. *Tape Notes*. Available from: <https://podcasts.apple.com/us/podcast/tn-144-bonobo/id1249834293?i=1000671480103> [Accessed 28 April, 2025]

Tape Notes podcast. (2024a). Episode 141, A.G. Cook. *Tape Notes*. Available from: <https://podcasts.apple.com/us/podcast/tape-notes/id1249834293?i=1000664535051> [Accessed 28 April, 2025]

Tape Notes podcast. (2024b). Episode 140, Four Tet. *Tape Notes*. Available from: <https://podcasts.apple.com/us/podcast/tn-140-four-tet/id1249834293?i=1000662459270> [Accessed 28 April, 2025]

Tape Notes podcast. (2024c). Episode 132, Jacob Collier. *Tape Notes*. Available from: <https://podcasts.apple.com/us/podcast/tn-132-jacob-collier/id1249834293?i=1000647602120> [Accessed 28 April, 2025]

Tape Notes podcast. (2024d). Episode 29, Floating Points. *Tape Notes*. Available from: <https://podcasts.apple.com/us/podcast/tn-29-floating-points/id1249834293?i=1000456635923> [Accessed 28 April, 2025]

Tape Notes podcast, (2023). Episode 116, Sub Focus. *Tape Notes*. Available from: <https://podcasts.apple.com/us/podcast/tn-116-sub-focus/id1249834293?i=1000617786889> [Accessed 28 April, 2025]

Tape Notes podcast. (2023a.) Episode 122, James Blake. *Tape Notes*. Available from: <https://podcasts.apple.com/us/podcast/tn-122-james-blake/id1249834293?i=1000627792182> [Accessed 28 April, 2025]

Tape Notes podcast. (2021). Episode 86, Jon Hopkins. *Tape Notes*. Available from: <https://podcasts.apple.com/us/podcast/tn-86-jon-hopkins/id1249834293?i=1000542794493> [Accessed 28 April, 2025]

Teibel, I. (1976). *Environments 7: Induced Meditation-a new easy method of relieving tension*. Syntonic Research Inc. SD 66007, LP.

Teibel, I. (1971). *Environments 3: totally new concepts in stereo sound*. Atlantic Records SC 66003, LP.

Telecom Electronic Beats TV. (2017). Feature: Skee Mask (Electronic Beats TV). *Telecom Electronic Beats*. Available from: <https://www.youtube.com/watch?v=jKtJKeBRKuY> [Accessed 28 April, 2025]

Théberge, P. (1997). *Any Sound You Can Imagine*. Middletown, Wesleyan University Press.

The Mr. Bill Podcast. (2025). Dave Gamble. Available from: <https://podcasts.apple.com/us/podcast/the-mr-bill-podcast/id1480159954?i=1000706795141> [Accessed 9 May, 2025]

The Mr. Bill Podcast. (2023). Barclay Crenshaw. Available from: <https://podcasts.apple.com/us/podcast/the-mr-bill-podcast/id1480159954?i=1000639328425> [Accessed 28 April, 2025]

The Mr. Bill Podcast. (2020). Steve Duda. Available from: <https://podcasts.apple.com/us/podcast/the-mr-bill-podcast/id1480159954?i=1000503420872> [Accessed 28 April, 2025]

Theise, N. (2023). *Notes On Complexity*. New York, Spiegel & Grau.

Thomson, P. (2024). Piano sketch to Orchestra with Spitfire Symphony Orchestra. *Paul Thomson*. Available from: <https://www.youtube.com/watch?v=dsdQI-Wcu6o> [Accessed 28 April, 2025]

Tie, T. (2022). How Fred Again. Transforms the Sounds of Social Media into Rave-Worthy Beauty. *NPR*. Available from: <https://www.npr.org/2022/11/12/1136054396/how-fred-again-transforms-the-sounds-of-social-media-into-rave-worthy-beauty> [Accessed 28 April, 2025]

Tingen, P. (2025). Inside Track: Koreless. *Sound On Sound*. Available from: <https://edition.soundonsound.com/html5/reader/production/default.aspx?pubname=&edid=c73acfeb-ed84-4bf2-ad65-891a1833c9b6> [Accessed 28 April, 2025]

Tingen, P. (1997). Harold Budd: American Vision. *Sound On Sound*. Available from: <https://www.soundonsound.com/people/harold-budd-american-vision> [Accessed 28 April, 2025]

Toffler, A. (1970). *Future Shock*. New York, Bantam.

Tone Glow. (2021). Seefeel. *Tone Glow*. Available from: <https://toneglow.substack.com/p/tone-glow-070-seefeel> [Accessed 28 April, 2025]

TossThisItem. (2022). *Why Is Ableton 'the One' That Everyone Uses. Reddit*. Available from: <www.reddit.com/r/TechnoProduction/comments/x5ulu9/why_is_ableton_the_one_that_everyone_uses/> [Accessed 28 April, 2025]

Trimbacher, P. et al. (2024). *The Secrets of Techno Production*. London, Attack Media Ltd.

Truax, B. (1996). Soundscape, Acoustic Communication and Environmental Sound Composition. *Contemporary Music Review*, 15 (1): 49–65.

Truax, B. (1978). *Handbook for Acoustic Ecology*. Vancouver, ARC Publications.

Truax, B., H. Westerkamp, and R.M. Schafer. (1973). *The Vancouver Soundscape 1973*. Cambridge Street Records.

Truss, S. (2024). 'The Prophet-6 Is My Most Used Synth. It's Got a Nice Balance of Not Trying to Do Too Much, but Doing Enough to Be Flexible': In the Studio with Max Cooper. *MusicRadar*. Available from: <https://www.musicradar.com/news/max-cooper-in-the-studio> [Accessed 28 April, 2025]

Truss, S. (2024a). It's Become Indispensable. *MusicRadar*. Available from: <https://www.musicradar.com/artists/producers-engineers/its-become-indispensable-i-use-it-on-almost-everything-because-you-simply-cant-go-wrong-with-it-mastering-engineer-and-musician-rafael-anton-irisarri-on-the-piece-of-gear-that-he-calls-his-secret-weapon> [Accessed 28 April, 2025]

Turner, D. (2023). Shapednoise: Accept No Compromise. *Ableton*. Available from: <https://www.ableton.com/en/blog/shapednoise-accept-no-compromise/?utm_source=2023-11-02-shapednoise&utm_medium=email&utm_term=l11-std-suite&utm_content=header&utm_campaign=editorial-newsletter&&sc_src=email_13259322&sc_lid=796428802&sc_uid=bTKBSpYVgP&sc_llid=67103&sc_eh=888a13d03e76b2851> [Accessed 28 April, 2025]

Turner, D. (2023a). Forest Swords: 'My First Couple of Records Were All Software - Then I Bought an Elektron Octatrack and It Blew My Mind Completely.' *MusicRadar*. Available from: <https://www.musicradar.com/news/forest-swords-interview> [Accessed 28 April, 2025]

Turner, D. (2024). Matmos on sampling dolphins, satellites, electromagnetic disturbances. *MusicRadar*. Available from: <https://www.musicradar.com/news/matmos-interview> [Accessed 28 April, 2025]

Turner, D. (2024a). Leo Abrahams: Rethinking Improvisation. *Ableton*. Available from: <https://www.ableton.com/en/blog/leo-abrahams-rethinking-improvisation/> [Accessed 28 April, 2025]

Turner, D. (2022). Objekt: 'The World Would Be a Boring Place Musically If Everyone Knew Exactly What They're Doing.' *MusicRadar*. Available from: <https://www.musicradar.com/news/objekt-interview-1> [Accessed 28 April, 2025]

Turner, L. (2013). Less Laptop, More Reality: Jon Hopkins Interviewed. *The Quietus*. Available from: <https://thequietus.com/interviews/jon-hopkins-interview/> [Accessed 28 April, 2025]

Twigger, R. (2018). *Micromastery: Learn Small, Learn Fast, and Unlock Your Potential to Achieve Anything*. New York, Tarcher.

Udvardyova, L. (2021). Julia Reidy: Sounding an Internal World. *SHAPE+ Platform*. Available from: <https://shape-platform.eu/2021/julia-reidy-sounding-an-internal-world/> [Accessed 28 April, 2025]

Veltman, C. (2023). Google's Latest AI Music Tool Creates Tracks Using Famous Singers' Voice Clones. *NPR*. Available from: <https://www.npr.org/2023/11/17/1213551049/googles-latest-ai-music-tool-creates-tracks-using-famous-singers-voice-clones> [Accessed 28 April, 2025]

Virtual Riot. (2024). Using AI to Make Music. *YouTube*. Available from: <https://www.youtube.com/watch?v=Hp05PlYEhR0&t=9051s> [Accessed 28 April, 2025]

Virtual Riot. (2023). Studio Time S2 EP3 - Track From Scratch. *YouTube*. Available from: <https://www.youtube.com/watch?v=kR2ubmupvQc&t=11s> [Accessed 28 April, 2025]

Virtual Riot. (2023a). Virtual Riot working w Skrillex, Origin of Color Bass, Sound Design Snugs, and Industry Pressure #10. *XLNTSOUND*. Available from: <https://www.youtube.com/watch?v=erZ9zwQvseM> [Accessed 28 April, 2025]

Voltage Labs. (2022). In the Studio with Legowelt / Voltage Labs. *YouTube*. Available from: <https://deref-gmx.com/mail/client/BQ15Y5HIkJs/dereferrer/?redirectUrl=https%3A%2F%2Fwww.youtube.com%2Fwatch%3Fv%3D3o_lxYImI7Y" \t "_blank"https://www.youtube.com/watch?v=3o_lxYImI7Y> [Accessed 14 July, 2025]

Walmsley, D. (2024). Radio/Live Transmission: Autechre as Concert Musicians. *The Quietus*. Available from: <https://thequietus.com/opinion-and-essays/black-sky-thinking/autechre-live-lp-radio-sessions-review/> [Accessed 28 April, 2025]

Walther-Hansen, M. (2020). *Making Sense of Recordings: How Cognitive Processing of Recorded Sound Works.* Oxford, Oxford University Press.

Ward, P. (2021). Classic Interview - Brian Eno. *MusicRadar.* Available from: <https://www.musicradar.com/news/brian-eno-interview> [Accessed 28 April, 2025]

Ward, P. (1993). Cagey, Canny, Krafty. *Music Technology*, July: 52–58.

Ware, T. (2007). Ricardo Villalobos: Sacred Art. *XLR8R.* Available from: <https://xlr8r.com/features/ricardo-villalobos-sacred-art/> [Accessed 28 April, 2025]

Warner Bros. (2022). The Sounds of Dune: Behind the Scenes with Denis Villeneuve. *Warner Bros. Entertainment.* Available from: <https://www.youtube.com/watch?v=Gkf_rdROyWw> [Accessed 28 April, 2025]

Watson, C. (2004). *Weather Report.* Touch Records: B00009U5MW.

Webb, D. (2020). *Dan Deacon Talks TM, Oblique Strategies and Synthesizers. Sungenre.* Available from: <https://sungenre.com/interview/dan-deacon/> [Accessed 28 April, 2025]

Weibel, P. (2014). Artist Interview: Ryoji Ikeda, creator of superposition. *University Musical Society.* Available from: <https://ums.org/2014/10/08/artist-interview-ryoji-ikeda-creator-of-superposition/> [Accessed 28 April, 2025]

Weisstein, E.W. (n.d.). Moiré Pattern. mathworld.wolfram.com. Available from: <https://mathworld.wolfram.com/MoirePattern.html> [Accessed 28 April, 2025]

Where Do I Find My Sound. (2022) *Reddit.* Available from: <https://www.reddit.com/r/edmproduction/comments/10acn30/where_do_i_find_my_sound/> [Accessed 28 April, 2025]

Willings, S. (2022). Flume: 'Being a Solo Producer Can Be Lonely. I Try to Make It Fun with as Low Stakes as Possible.' *MusicTech.* Available from: <https://musictech.com/features/interviews/flume-new-album-palaces/> [Accessed 28 April, 2025]

Westerkamp, H. (1974). Soundwalking. *Sound Heritage*, vol. 3, no. 4. Available from: <http://cec.sonus.ca/econtact/Soundwalk/Soundwalking.html> [Accessed 28 April, 2025]

Why We Bleep podcast. (2023). Rival Consoles, Episode 37. *Why We Bleep.* Available from: <https://podcasts.apple.com/us/podcast/why-we-bleep/id1338201154?i=1000626334854> [Accessed 28 April, 2025]

Why We Bleep podcast. (2020). Sarah Davachi, Episode 23. *Why We Bleep.* Available from: <https://podcasts.apple.com/us/podcast/why-we-bleep/id1338201154?i=1000496741010> [Accessed 28 April, 2025]

Wilson, C. (2010). 'I'll Be Bach': A Computer Program Is Writing Great, Original Works of Classical Music. Will Human Composers Soon Be Obsolete? *Slate.* Available from: <https://www.slate.com/articles/arts/music_box/2010/05/ill_be_bach.html> [Accessed 28 April, 2025]

WizBiz92. (2025). What are some intermediate to advanced audio only manipulation tips + tricks you use - and why?" *Reddit.* Available from: <https://www.reddit.com/r/ableton/comments/1jxeoit/what_are_some_intermediate_to_advanced_audio_only/?sort=old> [Accessed 28 April, 2025]

Wolfram, S. (2023). *What is ChatGPT Doing…And Why Does It Work?*Champaign, Wolfram Media.

Wong, M. (2024). AI Can't Make Music. *The Atlantic.* Available from: <https://www.theatlantic.com/technology/archive/2024/07/generative-ai-music-suno-udio/679114/> [Accessed 28 April, 2025]

Tie, T. (2022). How Fred Again. Transforms the Sounds of Social Media into Rave-Worthy Beauty. *NPR.* Available from: <https://www.npr.org/2022/11/12/1136054396/how-fred-again-transforms-the-sounds-of-social-media-into-rave-worthy-beauty> [Accessed 28 April, 2025]

Xenakis, I. (1992). *Formalized music: thought and mathematics in composition.* Hillsdale, Pendragon Press.

Xie, T. (2022). How Fred Again. transforms the sounds of social media into rave-worthy beauty. Available from: <https://www.npr.org/2022/11/12/1136054396/how-fred-again-transforms-the-sounds-of-social-media-into-rave-worthy-beauty" \t "_blank"https://www.npr.org/2022/11/12/1136054396/how-fred-again-transforms-the-sounds-of-social-media-into-rave-worthy-beauty> [Accessed 11 July 2025]

XLN Audio. (2025). Life by XLN Audio. Infinite Beats from Any Sound. *XLN Audio.* Available from: <https://www.xlnaudio.com/products/life> [Accessed 28 April, 2025]

XLR8R. (2017). Ask the Experts: Shifted. *XLR8R*. Available from: <https://xlr8r.com/features/ask-the-experts-shifted/> [Accessed 28 April, 2025]
XLR8R. (2018). In the Studio: DjRUM. *XLR8R*. Available from: <https://xlr8r.com/features/in-the-studio-djrum/> [Accessed 28 April, 2025]
XLR8R. (2017). Ask the Experts: Shifted. *XLR8R*. Available from: <https://xlr8r.com/features/ask-the-experts-shifted/> [Accessed 28 April, 2025]
Yanagi, Soetsu. (2013). *The Unknown Craftsman: A Japanese Insight into Beauty*. New York, Kodansha International.
Yim, K. (2023). Turn Ideas into Music with MusicLM. *MusicLM*. Available from: <https://blog.google/technology/ai/musiclm-google-ai-test-kitchen/> [Accessed 28 April, 2025]
Zanes, W. (2023). *Deliver Me from Nowhere: The Making of Bruce Springsteen's Nebraska*. New York, Crown.

INDEX

For Product Safety Concerns and Information please contact our EU
representative GPSR@taylorandfrancis.com
Taylor & Francis Verlag GmbH, Kaufingerstraße 24, 80331 München, Germany

www.ingramcontent.com/pod-product-compliance
Ingram Content Group UK Ltd.
Pitfield, Milton Keynes, MK11 3LW, UK
UKHW050229200926
475702UK00020B/2934
* 9 7 8 1 0 3 2 9 5 4 8 6 8 *